AF595028

The Advances and Applications of Optogenetics

The Advances and Applications of Optogenetics

Editors

Elena G. Govorunova
Oleg A. Sineshchekov

MDPI • Basel • Beijing • Wuhan • Barcelona • Belgrade • Manchester • Tokyo • Cluj • Tianjin

Editors
Elena G. Govorunova
Department of Biochemistry and Molecular Biology, University of Texas Medical School at Houston McGovern Medical School
USA

Oleg A. Sineshchekov
Department of Biochemistry and Molecular Biology, University of Texas Medical School at Houston McGovern Medical School
USA

Editorial Office
MDPI
St. Alban-Anlage 66
4052 Basel, Switzerland

This is a reprint of articles from the Special Issue published online in the open access journal *Applied Sciences* (ISSN 2076-3417) (available at: https://www.mdpi.com/journal/applsci/special_issues/Advances_Optogenetics).

For citation purposes, cite each article independently as indicated on the article page online and as indicated below:

LastName, A.A.; LastName, B.B.; LastName, C.C. Article Title. *Journal Name* **Year**, *Article Number*, Page Range.

ISBN 978-3-03943-673-6 (Hbk)
ISBN 978-3-03943-674-3 (PDF)

Cover image courtesy of Elena G. Govorunova.

Contents

About the Editors

Elena Govorunova holds a Ph.D. degree in Physiology from the M.V. Lomonosov Moscow State University in Moscow, Russia. Currently, she is a research associate professor at the Department of Biochemistry & Molecular Biology of the University of Texas Health Science Center at Houston McGovern Medical School, Houston, USA. Her research is focused on the identification and characterization of new channelrhodopsin variants for optogenetic applications, as well as the elucidation of the molecular mechanisms of ion conduction by these proteins. She has been a visiting researcher at the University of Cambridge (United Kingdom), University of Regensburg (Germany), Philipps-University Marburg (Germany) and Yale University (USA).

Oleg Sineshchekov holds a Ph.D. degree in Biophysics from the M.V. Lomonosov Moscow State University in Moscow, Russia. At present, he is a full research professor at the Department of Biochemistry & Molecular Biology of the University of Texas Health Science Center at Houston McGovern Medical School, Houston, USA. The focus of his research is mechanistic studies on microbial rhodopsins, for which he uses a variety of biophysical techniques, including a suspension assay for photoelectrical recording from algal suspensions that allows probing channelrhodopsins in their native cells. He has been a visiting researcher at the Biological Research Center in Szeged (Hungary), Philipps-University Marburg (Germany), Max-Planck Institute for Biochemistry in Martinsried (Germany), Worcester Foundation for Biomedical Research in Shrewsbury (USA) and the University of Tokyo (Japan).

Editorial

Editorial on Special Issue "The Advances and Applications of Optogenetics"

Elena G. Govorunova * and Oleg A. Sineshchekov

Center for Membrane Biology, Department of Biochemistry & Molecular Biology, McGovern Medical School, The University of Texas Health Science Center at Houston, Houston, TX 77030, USA; Oleg.A.Sineshchekov@uth.tmc.com

* Correspondence: Elena.G.Govorunova@uth.tmc.com

Received: 17 September 2020; Accepted: 18 September 2020; Published: 20 September 2020

Abstract: This Special Issue provides an update for the rapidly developing technology known as "optogenetics" that is the use of genetically encoded light-sensitive molecular elements (usually derived from lower organisms) to control or report various physiological and biochemical processes within the cell. Two ongoing clinical trials use optogenetic tools for vision restoration, and optogenetic strategies have been suggested as novel therapies for several neurological, psychiatric and cardiac disorders. The Special Issue comprises two reviews and seven experimental papers on different types of light-sensitive modules widely used in optogenetic studies. These papers demonstrate the efficiency and versatility of optogenetics and are expected to be equally relevant for advanced users and beginners who only consider using optogenetic tools in their research.

Keywords: optogenetics; photoswitching; photocontrol; all-optical electrophysiology; microbial rhodopsins; ion channels; LOV domains; membrane potential; intracellular trafficking; protein–protein interaction; signaling

---

1. Introduction

Broadly defined, optogenetic technology "combines genetic targeting of specific neurons or proteins with optical technology for imaging or control of the targets within intact, living neural circuits" [1]. This umbrella term encompasses both genetically encoded light-sensitive actuators and reporters of cellular activity. Historically, the reporters have been introduced first: targeting specific cell populations by heterologous expression of the gene encoding green fluorescent protein (GFP) from the jellyfish *Aequorea victoria* predated the term "optogenetics" by >10 years [2]. Structure-directed combinatorial mutagenesis of GFP has converted this protein into a fluorescent pH indicator to monitor synaptic transmission [3].

These early developments led Francis Crick to predict the possibility also to activate neurons with light [4]. Indeed, this has soon been achieved by co-expression of several essential elements of the enzymatic cascade of animal vision in non-visual cells [5]. However, the real coming of age optogenetics experienced after the emergence of a cornucopia of photosensitive molecules from photosynthetic microbes and plants. Furthermore, synthetic chromophores—referred to as "photoswitches"—have been designed to interact with specific target proteins and confer photosensitivity to them. Currently, many different natural and synthetic photosensitive moieties are being used in optogenetic experiments in many different cellular and organismal contexts, and the field is still rapidly expanding.

2. In This Special Issue

Piatkevich et al. review recent efforts to engineer genetically-encoded fluorescence indicators to monitor the membrane voltage and the concentrations of Ca^{2+} and K^{+}, as well as key neurotransmitters, changes in which accompany neuronal activity. This work serves as an excellent guide for selection of the most appropriate optogenetic reporters for a particular experiment.

Optogenetic actuators are even more diverse than sensors, in both their nature and intended uses. In some cases, such as microbial rhodopsins, the functions of the photosensor and effector are executed by the same protein domain, whereas in other proteins a photosensory domain is followed by distinct effector domains. Examples of photosensory domains found in native multidomain proteins are small flavoprotein modules known as Light, Oxygen, or Voltage sensing (LOV) and Blue-Light-Utilizing Flavin-binding (BLUF) domains that respond to UV-A/blue light (320–500 nm) [6,7]. Both these domains are widely used in optogenetic studies.

In plant phototropins that contain LOV domains, photoexcitation of the chromophore flavin mononucleotide (FMN) leads to unfolding of the C-terminal Jα helix, to which various peptides of interest, such as nuclear localization and export signals, can be attached. Wehler and di Ventura use a LOV domain-based light-inducible nuclear export system (LEXY) to manipulate cellular levels of the transcription factor p53 with blue light. In certain human cancers, excessive inactivation of p53 results from overexpression of its negative regulator, murine double minute 2 (Mdm2). The 12-amino-acid peptide, p53–Mdm2/MdmX inhibitor (PMI), binds to Mdm2 and suppresses its function. The authors show that in the dark, the PMI-LEXY fusion remains in the nucleus and prevents Mdm2 from degrading p53. Illumination caused export of the PMI-LEXY fusion to the cytosol, which released Mdm2. According to the authors, this optogenetic tool can be used to study the effects of local p53 activation within a tissue or organ.

BLUF domains are mostly found in prokaryotes and usually bind flavin adenine dinucleotide (FAD) as a chromophore. They exhibit different photochemical reactions, as compared to LOV domains. Kaushik et al. have analyzed 34 native BLUF domains from publicly accessible sequence databases. They have found functional association of these domains with several previously unknown effector domains, such as guanine nucleotide exchange factor for Rho/Rac/Cdc42-like GTPases (RhoGEF), phosphatidyl-ethanolamine binding protein (PBP), ankyrin and leucine-rich repeats. This remarkable modular diversity of BLUF domain-containing proteins expands the repertoire of potential chimeric assemblies that can be created by a combination of BLUF domains with appropriate cellular effectors.

Microbial rhodopsins, being electrogenic, are used to control the membrane voltage with light [8]. Channelrhodopsins mediate passive transport of ions along the electrochemical gradient and are therefore intrinsically more potent than rhodopsin ion pumps that translocate across the membrane only one ion per captured photon. Both cation- and anion-selective channelrhodopsins are known, abbreviated as CCRs and ACRs, respectively [9]. CCRs appear to emerge by convergent evolution by at least two independent routes. One CCR family was found in green (chlorophyte) flagellate algae, in which they act as photoreceptors for phototaxis [10]. Another CCR family that shows closer primary sequence homology to haloarchaeal rhodopsin pumps than to other known CCRs, was found in phylogenetically distant cryptophyte algae [11]. Shigemura et al. characterize channel properties of CCR4 from the cryptophyte alga *Guillardia theta* (GtCCR4). The advantages of this protein as an optogenetic tool comprise the red-shifted absorption maximum (530 nm), small desensitization during continuous illumination and the relatively high Na^+/H^+ permeability ratio, as compared to ChR2 from the chlorophyte alga *Chlamydomonas reinhardtii* (*Cr*ChR2).

H^+ permeability of CCRs is a serious problem in some optogenetic experiments, as it may lead to a decrease in the cytoplasmic pH [12]. Duan et al. show that the D156H mutation of *Cr*ChR2, and the corresponding mutations of the fast CCR variant Chronos [13] and blue-shifted ChR from *Platymonas subcordiformis* (*Ps*ChR) [14] enhanced relative permeabilities for Na^+ and Ca^{2+}, as compared to that for H^+. Moreover, in *Ps*ChR this mutation additionally increased the current amplitude, which made it the best currently available tool for optogenetic manipulation of the intracellular Ca^{2+} level.

Despite >50 native and the innumerable number of engineered CCR variants currently known, *Cr*ChR2 and its gain-of-function H134R mutant so far have remained the most frequently used optogenetic excitatory tools [15]. Two articles in this Special Issue report mechanistic studies on *Cr*ChR2, the results of which might contribute to further improvement of this tool for optogenetic needs. Richards et al. probe the role of residual hydrophobic mismatch (RHM) by a combination of computational and functional approaches. The authors identified several residues at the intracellular/lipid interface, mutations of which were predicted to significantly reduce the RHM energy penalty. They also showed, by electrophysiological analysis of these mutants, that the reduction of the RHM penalty in the closed state compromised *Cr*ChR2 conductance, selectivity and open state stability. These results show that protein–lipid interactions have to be taken into account when engineering optogenetic tools for specific cell types.

Ehrenberg et al. examine the functional role of Thr127 located near the retinylidene Schiff base in *Cr*ChR2. Replacement of this residue with alanine or serine did not change the position of the spectral maximum, which ruled out its contribution to the counterion complex. However, the T127A mutation, unlike the conservative T127S mutation, accelerated deprotonation of the Schiff base and strongly delayed its reprotonation. These results place Thr127 in the hydrogen-bonded network connecting the Schiff base with Asp156, which the authors identified earlier as the proton donor to the Schiff base [16]. This conclusion was further corroborated by the observation of extended lifetime of the channel open state observed in both T127A and T127S mutants, as compared to the wild type.

Erofeev et al. systematically analyzed the influence of frequency, duration and intensity of optical stimulation on performance of *Cr*ChR2 in cultured mouse hippocampal neurons. Using optimal photostimulation protocols is very important in optogenetic experiments, because e.g., insufficient illumination results in poor fidelity, whereas excessive light might lead to overheating of the tissue. The authors show that at the optimal stimulation frequency 1–5 Hz the dependence of photocurrent on the light pulse duration is described by a right-skewed bell-shaped curve, whereas the dependence on the stimulus intensity is close to linear. These results complement previously published work (e.g., [17]) and provide useful guidelines for optogenetic experimentation.

Finally, the review by Kellner and Berlin summarize recent progress in the development of synthetic azobenzene switches and their optimization for two-photon excitation (2PE). Azobenzene is the most popular chromophore used in synthetic optogenetics, owing to its high quantum yield, solubility in water and minimal photobleaching. Most importantly, under photoexcitation azobenzene undergoes a rapid, robust isomerization from the *trans* to *cis* conformation that can be harnessed to drive biologically relevant conformational changes in target proteins. 2PE allows using near-infrared (NIR) light that better penetrates biological tissue to activate optogenetic molecules and provides three-dimensional single cell-level spatial resolution. However, typical azobenzene-based switches exhibit poor absorption of NIR. The authors describe several strategies that have been used to increase the 2P-absorption cross section of azobenzene-based photoswitches without compromising the rate of their response or other useful properties.

Taken together, the papers in this Special Issue are a valuable contribution towards a better understanding of photochemistry and biophysics of optogenetic tools, which provides the guidelines for further engineering to improve their performance.

Author Contributions: Both authors have contributed to the writing and editing of this manuscript and agreed to publication of its final version. All authors have read and agreed to the published version of the manuscript.

Funding: This research received no external funding.

Conflicts of Interest: The authors declare no conflict of interest.

References

1. Deisseroth, K.; Feng, G.; Majewska, A.K.; Miesenböck, G.; Ting, A.; Schnitzer, M.J. Next-generation optical technologies for illuminating genetically targeted brain circuits. *J. Neurosci.* **2006**, *26*, 10380–10386. [CrossRef] [PubMed]
2. Chalfie, M.; Tu, Y.; Euskirchen, G.; Ward, W.W.; Prasher, D.C. Green fluorescent protein as a marker for gene expression. *Science* **1994**, *263*, 802–805. [CrossRef] [PubMed]

3. Miesenböck, G.; De Angelis, D.A.; Rothman, J.E. Visualizing secretion and synaptic transmission with pH-sensitive green fluorescent proteins. *Nature* **1998**, *394*, 192–195. [CrossRef] [PubMed]
4. Crick, F. The impact of molecular biology on neuroscience. *Philos. Trans. R. Soc. Lond. B Biol. Sci.* **1999**, *354*, 2021–2025. [CrossRef] [PubMed]
5. Zemelman, B.V.; Lee, G.A.; Ng, M.; Miesenböck, G. Selective photostimulation of genetically chARGed neurons. *Neuron* **2002**, *33*, 15–22. [CrossRef]
6. Christie, J.M.; Gawthorne, J.; Young, G.; Fraser, N.J.; Roe, A.J. LOV to BLUF: Flavoprotein contributions to the optogenetic toolkit. *Mol. Plant* **2012**, *5*, 533–544. [CrossRef] [PubMed]
7. Losi, A.; Gärtner, W. The evolution of flavin-binding photoreceptors: An ancient chromophore serving trendy blue-light sensors. *Annu. Rev. Plant Biol.* **2012**, *63*, 49–72. [CrossRef] [PubMed]
8. Zhang, F.; Vierock, J.; Yizhar, O.; Fenno, L.E.; Tsunoda, S.; Kianianmomeni, A.; Prigge, M.; Berndt, A.; Cushman, J.; Polle, J.; et al. The microbial opsin family of optogenetic tools. *Cell* **2011**, *147*, 1446–1457. [CrossRef] [PubMed]
9. Govorunova, E.G.; Sineshchekov, O.A.; Li, H.; Spudich, J.L. Microbial rhodopsins: Diversity, mechanisms, and optogenetic applications. *Annu. Rev. Biochem.* **2017**, *86*, 845–872. [CrossRef] [PubMed]
10. Sineshchekov, O.A.; Jung, K.-H.; Spudich, J.L. Two rhodopsins mediate phototaxis to low- and high-intensity light in *Chlamydomonas reinhardtii*. *Proc. Natl. Acad. Sci. USA* **2002**, *99*, 8689–8694. [CrossRef] [PubMed]
11. Govorunova, E.G.; Sineshchekov, O.A.; Spudich, J.L. Structurally distinct cation channelrhodopsins from cryptophyte algae. *Biophys. J.* **2016**, *110*, 2302–2304. [CrossRef] [PubMed]
12. Lin, J.Y.; Lin, M.Z.; Steinbach, P.; Tsien, R.Y. Characterization of engineered channelrhodopsin variants with improved properties and kinetics. *Biophys. J.* **2009**, *96*, 1803–1814. [CrossRef]
13. Klapoetke, N.C.; Murata, Y.; Kim, S.S.; Pulver, S.R.; Birdsey-Benson, A.; Cho, Y.K.; Morimoto, T.K.; Chuong, A.S.; Carpenter, E.J.; Tian, Z.; et al. Independent optical excitation of distinct neural populations. *Nat. Methods* **2014**, *11*, 338–346. [CrossRef] [PubMed]
14. Govorunova, E.G.; Sineshchekov, O.A.; Li, H.; Janz, R.; Spudich, J.L. Characterization of a highly efficient blue-shifted channelrhodopsin from the marine alga *Platymonas subcordiformis*. *J. Biol. Chem.* **2013**, *288*, 29911–29922. [CrossRef]
15. Wietek, J.; Prigge, M. Enhancing channelrhodopsins: An overview. *Methods Mol. Biol.* **2016**, *1408*, 141–165. [PubMed]
16. Lorenz-Fonfria, V.A.; Resler, T.; Krause, N.; Nack, M.; Gossing, M.; Fischer von Mollard, G.; Bamann, C.; Bamberg, E.; Schlesinger, R.; Heberle, J. Transient protonation changes in channelrhodopsin-2 and their relevance to channel gating. *Proc. Natl. Acad. Sci. USA* **2013**, *110*, E1273–E1281. [CrossRef] [PubMed]
17. Ishizuka, T.; Kakuda, M.; Araki, R.; Yawo, H. Kinetic evaluation of photosensitivity in genetically engineered neurons expressing green algae light-gated channels. *Neurosci. Res.* **2006**, *54*, 85–94. [CrossRef] [PubMed]

Review

Advances in Engineering and Application of Optogenetic Indicators for Neuroscience

Kiryl D. Piatkevich [1,*], Mitchell H. Murdock [1] and Fedor V. Subach [2,3,*]

1 Media Lab, McGovern Institute for Brain Research, and Department of Brain and Cognitive Sciences, MIT, Cambridge, MA 02139, USA; mitchm@mit.edu
2 INBICST, Moscow Institute of Physics and Technology, 123182 Moscow, Russia
3 Kurchatov Institute National Research Center, 123182 Moscow, Russia
* Correspondence: kiryl.piatkevich@gmail.com (K.D.P.); subach.fv@mipt.ru or subach_fv@nrcki.ru (F.V.S.)

Received: 15 January 2019; Accepted: 2 February 2019; Published: 8 February 2019

Abstract: Our ability to investigate the brain is limited by available technologies that can record biological processes in vivo with suitable spatiotemporal resolution. Advances in optogenetics now enable optical recording and perturbation of central physiological processes within the intact brains of model organisms. By monitoring key signaling molecules noninvasively, we can better appreciate how information is processed and integrated within intact circuits. In this review, we describe recent efforts engineering genetically-encoded fluorescence indicators to monitor neuronal activity. We summarize recent advances of sensors for calcium, potassium, voltage, and select neurotransmitters, focusing on their molecular design, properties, and current limitations. We also highlight impressive applications of these sensors in neuroscience research. We adopt the view that advances in sensor engineering will yield enduring insights on systems neuroscience. Neuroscientists are eager to adopt suitable tools for imaging neural activity in vivo, making this a golden age for engineering optogenetic indicators.

Keywords: optogenetic tools; neuroscience; calcium sensor; voltage sensor; neurotransmitters

---

1. Introduction

A central and astonishing feature of the nervous system is the capacity for learning and remembering, which are inherently dynamic processes. Advances in genetically-encoded sensors enable the real-time observation of key signaling molecules in a cell-type and circuit-specific manner within intact brain tissue and in vivo. Genetic strategies allow targeted expression of optogenetic tools [1,2] to a specific cell type (using specific promoters [3] or in transgenic animals expressing DNA recombinases in a specific cell type [4]) or an anatomically distinct circuit (using intersectional or retrograde viral labeling strategies [5,6]). Additionally, genetically-encoded probes are the only technique available to observe precisely the same cells longitudinally, permitting long-term monitoring of specific cellular processes, up to months [7,8]. Judiciously selecting suitable spectral properties of optical sensors potentially enables the visualization of the activity and interactions of distinct cell types [9]. Thus, genetically-encoded indicators are an indispensable tool for visualizing neuronal activity in a cell type and circuit-specific manner while minimizing disturbance to the complex cellular milieu of the brain. Advances in these sensors allow noninvasive and longitudinal monitoring of key neural processes, which is essential for understanding how information is transmitted and processed.

In this review, we describe advances from the past three years in the engineering and application of genetically-encoded fluorescence indicators of neuronal activity. We focus on indicators for calcium, voltage, and neurotransmitters that show acceptable performance for in vivo imaging. We also provide an overview of new sensors—which potentially enable fundamentally new kinds of measurements—and their molecular design, biochemical and spectral characteristics, and current limitations.

2. Calcium Indicators

Calcium is a crucial mediator of neural activity and activity-dependent synaptic plasticity [10,11]. While most neurons at rest contain cytoplasmic calcium concentrations of 50–100 nM, electrical activity can swiftly and dramatically increase calcium concentrations [12] for tens of milliseconds to several seconds [13,14]. Therefore, calcium is an excellent proxy for neuronal activity, and, accordingly, genetically-encoded calcium indicators (GECIs) are the most popular and widely used optical sensors of neuronal activity in neuroscience [8,11,15,16]. Since the development of the first proof of principle GECIs almost two decades ago [17–20], herculean efforts in protein engineering resulted in several excellent calcium sensors [21,22] providing researchers a selection of tools for diverse applications, including imaging large neural population dynamics [23,24], dendritic processing [25], as well as synaptic [26,27] and presynaptic [28] function. Here, we will discuss the major progress in engineering GECIs from the past few years and briefly describe some of their most distinctive applications. Prior work on the development and application of GECIs has been reviewed in earlier publications [8,11,15,16].

Since 2016, two types of molecular designs have prevailed among improved as well as newly developed GECIs (Figure 1a,b). The GECI families based on the GCaMP-like design are the most numerous and widely used GECIs for in vivo imaging [22] (Table 1). Calcium sensors with green fluorescence exhibit the best performance, especially the GCaMP6 variants [26] and, therefore, are the primary choice for most applications. Intensive use of the GCaMP variants in vivo resulted in a wealth of evidence for their interference with normal calcium dynamics and gene expression in mammalian systems that must be taken into account when interpreting calcium imaging data [29–31]. Typically, the side effects are due to the interaction of the calcium binding domain, which has a mammalian origin, with endogenous proteins, as well as the buffering of cytoplasmic calcium, artifacts more prominent during prolonged sensor expression at high levels [26,32]. To overcome these side effects, several different modifications of calcium binding domains have been attempted. For example, Yang et al. engineered GCaMP-X by incorporating an extra apoCaM-binding motif into the GCaMP variants [31]. This modification did not significantly affect overall sensor performance but reduced interactions with L-type calcium channels, thus effectively protecting Ca_V1-dependent excitation-transcription coupling from sensor-induced perturbations [31]. An alternative strategy to minimize GCaMP-induced side effects involve exploiting calcium binding domains cloned from fungi and yeasts that share conserved amino acid identity with their metazoan counterparts used in GCaMPs [33]. For example, the calmodulin and M13-like peptide from *Aspergillus* fungi swapped with the calcium-binding domain in GCaMP6s prevented interaction with endogenous proteins at low calcium concentrations in cultured mammalian cells [33]. In addition, fungal GCaMP, or FGCaMP, exhibits ratiometric by excitation fluorescence response to calcium ions, with the highest brightness and dynamic range combination among other ratiometric GECIs such as GEX-GECO [34], Pericam [19], and Y-GECO [35]. Furthermore, FGCaMP can report neuronal activity with single cell resolution in zebrafish larvae under light-sheet microscopy. Another way to improve the performance of GECIs is structure-guided mutagenesis of the GFP-CaM interface and the CaM-M13 peptide interactions [36]. For example, further mutagenesis of the calcium binding domain in the GCaMP6 variants resulted in the next generation green sensors, the jGCaMP7 series, characterized by improved sensitivity to one action potential and higher signal-to-noise ratio due to enhanced brightness [37]. However, the jGCaMP7 variants are still less validated in vivo compared to the extremely popular GCaMP6 indicators.

Very recently, Barykina et al. suggested an alternative design of GECIs, which implicates the insertion of a calcium-binding domain into a fluorescent protein (Figure 1b). Implementing this design resulted in the generation of a new family of GECIs, named after the progenitor NTnC [38], exhibiting a set of unique features (Table 1). In comparison to GECIs with the classical GCaMP design, the NTnC-like family is characterized by a smaller molecular size, lower calcium-binding capacity, higher tolerance to fusion partners, and a non-mammalian origin of calcium-binding domains. The smaller molecular size is beneficial for packaging efficiency into viral particles and perhaps ensures better folding and targeting to subcellular compartments, such as mitochondria and the endoplasmic reticulum [39]. The lower calcium-binding capacity of the NTnC family (two or one calcium ion per molecule vs four ions per molecule for GCaMP family) reduces the deleterious confound of calcium buffering, which can corrupt the patterns of registered neuronal activity [30]. However, in spite of the different stoichiometry of calcium binding sites in GCaMP and NTnC-like indicators, they linearly respond to the increasing number of action potentials in the range determined by affinity of the respective indicator to calcium ions [38,39]. Unlike the GCaMP-like GECIs, where different tags affect their dynamic range and affinity for calcium ions [40], the GFP-like N- and C-termini make the NTnC-like sensors tolerate fusions with other proteins by eliminating distortion of calcium-binding domain. Utilization of the truncated version of troponin C from muscles as a calcium-binding domain, which does not interact with endogenous proteins in mammalian cells, minimizes potential disturbances on neural physiology.

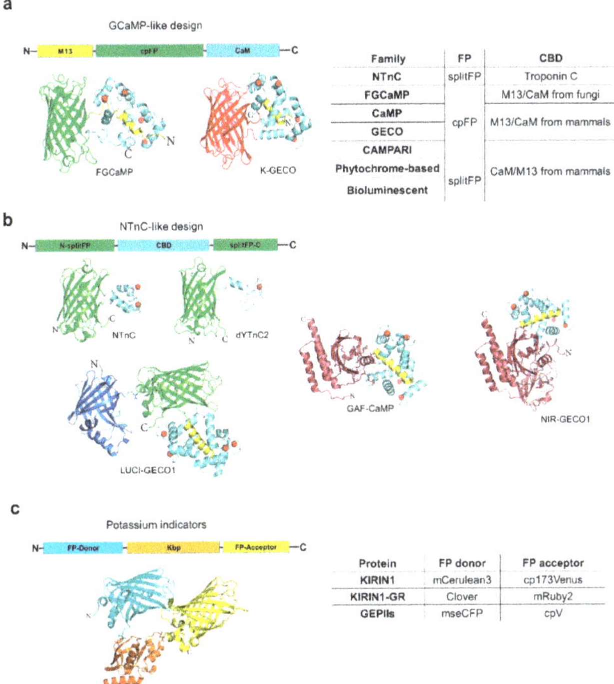

Figure 1. Molecular design of genetically-encoded calcium and potassium indicators. (**a**,**b**) The structure of calcium indicators are shown as ribbon diagrams according to the crystal structures of NTnC (protein database (PDB) 5MWC), dYTnC2 (unpublished), GAF-CaMP (combination of GAF-domain of PaBphP phytochrome PDB 3C2W and CaM/M13-peptide pair from GCaMP6m GECI PDB 3WLD), NIR-GECO1 (combination of the PAS-GAF domain of PaBphP phytochrome PDB 3C2W and CaM/M13-peptide pair from GCaMP6m GECI PDB 3WLD), LUCI-GECO1 (combination of NanoLuc luciferase PDB 5IBO and GCaMP6m GECI PDB 3WLD), FGCaMP (unpublished), and K-GECO (PDB 5UKG) in calcium-bound state. (**c**) The organization of potassium indicators is shown schematically and according to the X-ray structures of Cerulean and mVenus fluorescent proteins (PDB 5OXC and 1MYW, respectively), and Kbp potassium binding protein from *E.coli* (PDB 5FIM).

Table 1. The key characteristics and performance of the selected genetically-encoded fluorescent calcium and potassium indicators.

Family	Indicator [a]		Ex/Em (nm) [b]	Brightness vs. EGFP (%) [c]	$pK_{a,apo}/pK_{a,sat}$	ΔF/F (%) [d]	K_d (nM) [e]	k_{off} (s^{-1}) [f]	ΔF/F/SNR per AP		Ref.
									Slice or Culture	*In Vivo*	
Calcium indicators											
NTnC	NTnC		505/518	163	6.09/6.08	100	192	0.8	0.027/18	ND	[38]
	iYTnC2		499/518	16	7.4/8.5	450	331	1.12	0.0015/ND	ND	[41]
	YTnC		495/516	17	5.2/6.3	290	410	0.96	0.008/ND	ND	[39]
	dYTnC2		496/518	ND	ND	1900	2700	ND	ND	ND	[unpublished]
FGCaMP	FGCaMP®	apo	402/516	56	6.56/7.0	590	400	1.2	ND	ND	[33]
		sat	493/516	104	6.2/7.33	1370	460, 4400				
CaMP	GCaMP6s		497/515	124	9.77/6.20	6220	227	0.69-1.12	0.022–0.28/7.6–183	0.25/ND	[26]
	GCaMP6f		497/515	109	8.77/6.34	5080	492	3.93	0.18/101	0.15/ND	[26]
	jGCAMP7s		497/515	103	7.69/6.36	3900	68 *	2.86	0.657/25	ND	[37]
GECO	jRGECO1a		560/590	35	8.6/6.3	1060	148 *	7.6	−0.29/−13	0.13/ND	[9,42]
	K-GECO1		565/590	82	ND	1100	165 *	ND	−0.26/−11	ND	[43]
CAMPARI	CAMPARI2	G	502/516	268	ND	7.8	199 *	1.43	ND	ND	[40]
		R	562/577	126	ND						
Phytochrome-based	GAF-CaMP		635/672	21, 69 *	3.0; 9.0/ 3.0; >10.0	52	15	ND	ND	ND	[44]
	NIR-GECO1		678/704	ND	6.0/4.7	85	215	1.93	2.5/25	ND	[45]
Bioluminecent	LUCI-GECO1[BRET]	donor	460	ND	8.15/6.07	406	285 *	ND	ND	ND	[46]
		acceptor	497/515								
Potassium indicators											
GEPI	KIRIN1[FRET]	donor	433/475	104	3.2	130	1660	ND	ND	ND	[47]
		acceptor	515/530	ND	ND						
	KIRIN1-GR[FRET]	donor	505/515	251	6.2	20	2560	ND	ND	ND	
		acceptor	559/600	128	5.3						
	GEPII[FRET]	donor	457/475	ND	ND	220	420–26,000	0.75	ND	ND	[48]
		acceptor	515/527								
	GINKO1		502/514	22	ND	250	420	ND	ND	ND	[47]

[a] ®, ratiometric, [FRET]—Förster (fluorescence) resonance energy transfer (FRET)-based indicator, [BRET]—bioluminescence resonance energy transfer (BRET)-based indicator. [b] Excitation/emission wavelengths for the brightest state. For ratiometric and FRET-based indicators two wavelengths or two pairs of wavelengths correspond to apo- and saturated-states or ex/em of the donor and acceptor, respectively. [c] Brightness is a product of quantum yield and extinction coefficient normalized to the brightness of EGFP. For ratiometric and FRET-based indicators brightness of apo- and saturated-states or fluorescent protein acceptor and donor are shown, respectively, *—two-photon brightness. [d] Dynamic range is maximally achievable ΔF/F between calcium/potassium-saturated and apo-states. [e] K_d is the equilibrium calcium dissociation constant in the presence of 1mM $MgCl_2$, * K_d values measured in the absence of 1mM $MgCl_2$; [f] k_{off} are the off kinetics of dissociation from calcium/potassium ions measured using stopped flow fluorimetry. ND—not determined.

The original NTnC sensor was designed by inserting the troponin C calcium binding domain into the bright green fluorescent protein mNeonGreen [38]. In contrast to GCaMP-like sensors, NTnC exhibits negative fluorescence response, i.e., it reduces fluorescence upon calcium binding. NTnC is characterized by high brightness and pH stability, inherited from mNeonGreen, but low dynamic range (Table 1). Utilizing yellow fluorescent protein (YFP) as a sensing moiety helped to increase the dynamic range but at the expense of brightness and pH stability [39,41]. While NTnC and YTnC are suitable for reporting neuronal activity at single cell resolution in behaving mice under both one- and two-photon microscopy, they exhibit lower overall performance than the GCaMP6 variants. However, we found the NTnC-like sensors significantly outperform GCaMP6s when targeted to organelles such as the mitochondria and endoplasmic reticulum [39]. Calcium homeostasis in mitochondria and the endoplasmic reticulum plays crucial roles in cell physiology in health and disease [49–52]. However, currently available GECIs are not optimized for calcium visualization in these organelles. Therefore, the NTnC-like sensors represent promising templates for adjusting calcium affinity to match the range of calcium concentration in mitochondria and the endoplasmic reticulum. For example, truncation of the EF4-hand of the calcium-binding domain in YTnC2 indicator, which is capable of binding only one calcium ion, decreased affinity of the indicator to calcium ions by ~10-fold. The generated sensor, dYTnC2, showed optimal biochemical characteristics for measuring large calcium transients in the endoplasmic reticulum (personal communication, Table 1).

There is a great need for GECIs with red-shifted fluorescence, but engineering red-shifted variants has proven particularly challenging. Inserting calcium binding domains into red fluorescent proteins easily disrupts folding and chromophore maturation [43,53]. Furthermore, red fluorescent proteins are known to exhibit inferior photophysical properties compared to green fluorescent proteins such as photoactivation under blue light illumination, leading to imaging artifacts [54–56]. Despite great efforts in the development of red-shifted GECIs [34,42,54,56–58], only the latest generation of red GECIs is suitable for imaging calcium dynamics in living organisms [9,43]. However, further engineering and enhancement of the current red GECIs is certainly required before they can reach level of the best performing green sensors (Table 1). Brightness and dynamic range are perhaps the main properties requiring improvement. Another inherent drawback of red GECIs is their tendency to form bright fluorescence puncta in cell bodies, especially during in vivo expression. The puncta were shown to co-localize with a lysosomal marker LAMP-1. The aggregated proteins in the puncta do not show calcium sensitivity and contribute to background fluorescence, thus reducing the overall quality of the dynamic signal recordings. To reduce puncta formation, Shen et al. exploited the novel red fluorescent protein FusionRed, known to exhibit good localization in neurons [43]. Replacing the fluorescent moiety in R-GECO with a circularly permutated FusionRed generated the K-GECO indicator, which indeed demonstrated no puncta-like localization in cultured neurons [43]. However, puncta formation was not completely resolved in vivo in mice. K-GECO could report single action potentials in neurons in zebrafish larva as well as detect calcium dynamics in the visual cortex of awake mice. However, K-GECO1 did not provide the same level of in vivo sensitivity as the highly optimized jRGECO1a (Table 1).

Red-shifted GECIs enable facile application in conjugation with channelrhodopsins for all-optical interrogation of neuronal circuits [9,42,43,56]. Due to the wide action spectra of the majority of channelrhodopsins, compared with the full width at half maximum of GFP-like fluorescent proteins ranging from 30 to 70 nm, spectrally multiplexed optogenetic control with red GECIs excited at ~560 nm still remains challenging. Even one of the most blue-shifted channelrhodopsins, CheRiff [59], with acceptable in vivo performance [60], generates a sustained inward current under 560 nm illumination at moderate light power of ~15mW/mm^2 that can cause substantial alternations in membrane potentials during extended imaging [61,62]. Shifting fluorescence of GECIs into the near-infrared range will provide completely crosstalk-free coupling with optogenetic tools. Near-infrared fluorescent proteins derived from phytochomes [63] have shown to be useful for labeling neurons in vivo in mammals [64], zebrafish [65], and flies [66]. Near-infrared light is also beneficial for the noninvasive whole-body imaging of small mammals [67,68] due to reduced autofluorescence, low light scattering, and minimal absorbance at longer wavelengths. Interestingly, the first attempts engineering calcium sensors based on the phytochrome-derived fluorescence proteins were unsuccessful. Obtaining functional circularly permutated variants was impossible due to the phytochrome's topology, with the N- and C-termini far apart from each other, unlike in GFP-like proteins (personal communications). Instead, the first near-infrared GECIs were engineered by inserting the calcium binding domain into phytochrome-derived fluorescent proteins. In this respect, their design appears similar to that of the NTnC-like family (Figure 1b). The GAF-FP [69] and mIFP [66] proteins were used as fluorescent scaffolds producing GAF-CaMP [44] and NIR-GECO1 [45] indicators containing cGMP phosphodiesterase/adenylate cyclase/FhlA transcriptional activator (GAF) and Per-ARNT-Sim-GAF (PAS-GAF) domains from the phytochromes, respectively [63]. While GAF-CaMP has not been yet tested in neurons, NIR-GECO1 was shown to detect single action potentials in cultured neurons and intact mouse brain tissue. NIR-GECO1 is efficiently excited by far-red light in the range from 630 to 680 nm, enabling crosstalk free spectral multiplexing with the majority of channelrhodopsins and other fluorescence sensors. In addition, NIR-GECO1 allowed mesoscale fluorescence imaging of somatosensory cortex through the intact skin and skull of anaesthetized mice, although reliable detection of fluorescent changes required averaging of multiple trials. As a first-generation indicator, NIR-GECO1 falls short of the most extensively optimized fluorescent protein-based GECIs in several critical performance parameters but is an exciting progression in the development of far red-shifted calcium sensors.

An alternative method to record neuronal activity of a large number of neurons with single cell resolution is the permanent labeling of neurons, as demonstrated by the CaMPARI calcium sensor [70]. The CaMPARI family of light-inducible calcium sensors is based on EosFP green-to-red photoconvertable fluorescent protein, allowing ex vivo visualization of integrated over time calcium transients (Figure 1). The CaMPARI integrators are irreversibly and efficiently photoconverted from a green fluorescent state to a red state by ultraviolet light (~405 nm) irradiation only in the presence of elevated concentrations of calcium ions. The second generation of this family, CaMPARI2, is characterized by improved dynamic range of photoconversion, brightness, faster calcium dissociation kinetics, decreased photoconversion rate in inactive neurons, and compatibility with paraformaldehyde fixation [40]. In addition, the authors introduced the anti-CaMPARI-red antibody that specifically recognizes the photoconverted red form and thus allows immunohistochemical amplification of calcium signal in fixed cells or tissue. The CaMPARI2 indicator enables acquisitions of snapshots of calcium as a proxy for neuronal activity integrated during 30 seconds in the brain of zebrafish from up to 6000 individual neurons. In this respect CaMPARI integrators are comparable with CaMP and GECO indicators which allow simultaneous two-photon imaging of calcium activity over 10,000 cells in visual cortex of mice with 2.5 Hz rate [71]. In spite of great progress in the development of the CaMPARI sensors, they provide only a single snapshot of neuronal activity acquired during tens of seconds and require irradiation with highly phototoxic ultraviolet light, which also suffers from poor depth penetration in live tissue.

A significant step forward was made for improving the class of calcium sensors that employs bioluminescence resonance energy transfer (BRET) phenomenon [72]. The newly-developed BRET-based LUCI-GECO1 indicator combines the bright NanoLuc luciferase with emission maximum at 460 nm with the topological version of GCaMP6s indicator (ncpGCaMP6s; Figure 1b) [46]. In comparison to the previously developed BRET-based GECIs, LUCI-GECO1 was characterized by slightly wider dynamic range and higher brightness. Due to the lack of need for illumination for fluorescence excitation, LUCI-GECO1 enables straightforward multiplexing with optogenetic tools, such as channelrhodopsins. While BRET-based calcium sensors hold great promise for noninvasive imaging in mammals, their application in vivo has not yet been demonstrated.

3. Potassium Indicators

Sodium and potassium are important physiological ions crucial in maintaining and regulating membrane potentials in neurons. In neurons at rest, cytoplasmic sodium levels are an order of magnitude lower than extracellular (~10 mM intracellular vs ~150 mM extracellular) [73]. During an action potential, intracellular sodium levels transiently increase [74,75], and because sodium is the major current carrier of glutamate-gated voltage changes, noninvasive optical imaging of sodium would enable the visualization of a key correlate of excitatory synaptic transmission. Potassium is also a crucial regulator of cell excitability [76,77]. At rest, cytoplasmic potassium levels are high, but during an action potential, there is a potassium efflux, the change in the absolute number of ions is relatively modest [76,77]. Noninvasively recording potassium at single cell resolution would offer an important proxy to neural activity. Unfortunately, there are no genetically-encoded sodium sensors developed to date, and genetically-encoded potassium sensors are only an emerging class of optical probes. We note that an important part of the difficulty in engineering potassium and sodium sensors—compared to calcium sensors—is that while calcium concentrations during an action potential change by more than four orders of magnitude, changes in potassium and sodium concentrations during an action potential are only one order of magnitude. Thus, imaging potassium and sodium is inherently more difficult than imaging calcium transients owing to the more subtle changes in ion concentration.

Genetically-encoded potassium indicators (GEPI) are currently represented mainly by three Förster (fluorescence) resonance energy transfer (FRET)-based indicators (Figure 1c; Table 1). All three GEPIs share the same Kbp protein from *E. coli* as a potassium binding domain and utilize CFP/YFP fluorescent FRET-pairs to register conformational changes in the Kbp part [47,48]. Since GEPI dissociation kinetics from potassium ions is limited by hundred milliseconds, they are not appropriate for the detection of fast potassium dynamics during neuronal activity. Meanwhile, GEPIs were successfully applied to monitor slower potassium transients in different cellular compartments in cultured cells. Thus, further development of faster and non-FRET-based variants of GEPIs with higher dynamic range are in high demand.

4. Voltage Sensors

Voltage changes are the primary means of millisecond-scale computation in neurons because they enable sophisticated fast dendritic computations and rapid transmission of signals across long distances [78]. Voltage changes are the most direct measure of a neuron's electrical potential [79]. Therefore, optical voltage sensors are an ideal method to noninvasively measure electrical activity at single cell resolution. However, despite recent progress in development of new voltage sensitive probes [60,65,80–82], voltage imaging still remains technically more challenging than calcium imaging. One of the major challenges is associated with the sub-millisecond timescale of membrane potential changes during neuronal activity. Subsequently, voltage imaging requires kHz acquisition rates to resolve brief voltage changes upon action potentials, while calcium imaging is typically performed at 1 to 10 Hz. Although currently available scientific-grade cameras can perform imaging at kHz rates, these high speed result in pixel counts reduced by an order of magnitude than those commonly used for calcium imaging. Furthermore, compared to the GECIs expressed in the bulk cytosol, the maximum number of voltage sensor molecules that can be localized on plasma membrane is several orders of magnitude smaller. These factors significantly reduce the number of photons integrated during each acquired frame making fluorescence signal increasingly difficult to distinguish cells from background fluorescence in brain tissue.

The first proof of principle of a fully genetically-encoded voltage sensor was reported in 1997 [83], followed by almost two decades of intensive protein engineering efforts to develop a fluorescence sensor capable of reporting neuronal activity with single cell, single spike resolution in awake mice [84]. To date a number of improved and newly engineered genetically-encoded voltage indicators (GEVIs) with diverse spectral and biophysical characteristics reached the level of performance sufficient for in vivo imaging with high temporal precision at single-cell resolution in awake mice surpassing beyond "proof of principle" applications (Table 2). Currently, the majority of developed GEVIs are based on two types of voltage-sensitive moieties: microbial rhodopsins and voltage-sensing domains (VSDs) from voltage-sensitive phosphatases. Naturally-occurring and engineered microbial rhodopsins have already found widespread application in neuroscience due to their ability to optically control electrical activity of genetically targeted neurons [85]. Photoactivated rhodopsin translocates ions across the plasma membrane thus modulating membrane potentials, usually in a predictable manner [86–88]. Several years ago, Adam Cohen's group made a discovery that paved the way for the utilization of opsins not only for manipulating membrane potential but also for observing changes in voltage across the plasma membrane. His group showed that one class of opsins, light-driven proton pumps, can work in reverse, i.e., emitting light in response to membrane voltage alternations [89,90]. Indeed, spectroscopic studies of the proton pumping mechanism in rhodopsins, cloned from archaea and bacteria, revealed a complex spectroscopic photocycle involving multiple intermediates with diverse absorption spectra [89–91]. Certain transitions between intermediates appeared to be voltage dependent while occurring at submillisecond time scales [92]. Furthermore, a single point mutation in the chromophore vicinity of proton pumps was enough to almost completely abolish light-driven currents while retaining their voltage dependent spectral changes. Altogether, these features confer optical recording of membrane potential dynamics at biologically-relevant temporal resolution via probing opsin absorption spectrum. Voltage modulated spectral changes, or the so called electrochromic effect, can be read-out via nonradiative quenching of an attached fluorescent moiety, a form of FRET [93,94]. Using electrochromic FRET (eFRET) approach, a palette of GEVIs with fluorescence ranging from green to red were engineered by appending various fluorescent proteins to the C-terminus of the proton pumping microbial opsins [80,84,95,96] (Figure 2a). The first generation of eFRET-opsin voltage sensors, incorporating Mac and Arch proton pumps [97] as voltage-sensing scaffolds, can report single action potentials in cultured neurons with a moderate signal-to-noise ratio (7–9 at 1 kHz acquisition rate) [95,96]. However, these sensors suffer from poor membrane localization and relatively slow kinetics, on the order of few milliseconds, that precluded their adaptation by the neuroscience community. Substantial improvement in overall performance was achieved by utilizing the blue-shifted *Acetabularia acetabulum* rhodopopsin (Ace) [98] in fusion with the very bright green fluorescent protein mNeonGreen. The resulting group of GEVIs, usually collectively called Ace2N-mNeon (2N stands for the point mutation abolishing proton pumping), is characterized by high brightness, moderate photostability, and submillisecond kinetics, providing high spike detection fidelity under wide-field microscopy in both ex vivo and in vivo preparations [84]. For example, the Ace2N-4aa-mNeon variant, containing four amino acids in the linker between Ace and mNeonGreen, was demonstrated to report single action potentials in individual neurons in the primary visual cortex of awake mice (Table 1). In turn, a variant with a shorter linker, Ace2N-2aa-mNeon, was able to detect odor-evoked voltage signals in single olfactory projection neurons and their dendritic arbors and axonal boutons in live fruit flies.

Table 2. Performance of selected genetically-encoded and hybrid fluorescent voltage sensors.

Sensor	Ex/Em (nm)	$\Delta F/F_{AP}$ (%)	SNR	On Kinetics [a] (ms)	Off Kinetics [a] (ms)	Number of Simultaneously Imaged Neurons In Vivo	Ref
Ace2N-4aa-mNeon	506/517	−6.5	ND	0.37 (58)/5.5	0.5 (60)/5.9	2 in primary visual cortex L2/3	[84]
ASAP3	485/510	−10 [b]	19 [b]	0.94 (72)/7.2	3.8 (76)/16	1 in primary visual cortex L1 and L2/3	[81]
Voltron$_{525}$	525/549	−6.5 [b]	27 [b]	0.64 (61)/4.1	0.78 (55)/3.9	46 in primary visual cortex L1	[82]
VARNAM	558/592	−4.8	12	0.9/5.2	0.8/4.7	Aggregated population signal	[80]
QuasAr2	590/715	15	8.5	0.3 (62)/3.2	0.3/(73) 4.0	14 in mouse nodose ganglia	[99]
paQuasAr3	590/715	23 [b]	28 [b]	0.9 (57)/15	0.93 (79)/15	6 in hippocampus	[60]
Archon1	590/715	53 [b]	37 [b]	0.61 (88)/8.1 [c]	1.1 (88)/13 [c]	18 in CA1 hippocampus	[65]

Ex—excitation wavelength; Em—emission wavelength; SNR—signal-to-noise ratio per single action potential measured in intact mouse brain slice at 22 °C under wide-field microscopy at 500 Hz acquisition rate; ND—not determined. [a] voltage kinetics evaluated by bi-exponential fitting (in the format fast/slow), where the value in parentheses represents the % of current magnitude in the fast τ component, measured in HEK cells at 34 °C; [b] values measured for soma localized version of the sensor; [c] measured in cultured neurons.

Coupling Ace2N with the bright fluorescent protein mRuby3 led to the development of a red-shifted GEVI named VARNAM [80], which inherited the fast kinetics of Ace2N and high brightness and photostability of mRuby3. However, despite a single point mutation that presumably improved eFRET efficiency by red-shifting absorbance spectrum of Ace2N, VARNAM exhibits lower voltage sensitivity than other advanced GENIs suitable for in vivo imaging (Table 1). For example, averaging up to 10 optical trails was required to resolve evoked excitatory and inhibitory postsynaptic potentials using VARNAM in intact brain tissue [80]. In live mice, VARNAM was able to track cell-type specific oscillatory dynamics for subsets of neurons in somatosensory and frontal cortex and CA1 of the hippocampus using the fiberoptic imaging. In addition, VARNAM reported spontaneous neuronal activity with single spike precision in multiple cell types in live fruit flies.

Another recently developed fluorescent voltage sensor utilizing Ace2N as a voltage sensing domain is Voltron [82]. While Voltron is not a fully genetically-encoded probe, but rather a chemogenetic or hybrid voltage sensor, we discuss it here in comparison with other GEVIs due to its high performance in multiple species in vivo, including mice, zebrafish, and fruit flies. Voltron combines Ace2N with a self-labeling protein tag domain that covalently binds improved rhodamine dyes such as the Janelia Fluor® (JF) dyes [100] (Figure 2b). While a palette of JF dyes with emission maxima ranging from 505 to 635 nm were shown to work well with Voltron in cultured neurons, the Voltron/JF525 combination exhibited the highest voltage sensitivity and, thus, used for all further in vivo applications. To perform voltage imaging in vivo in mice, Voltron was fused to the soma-localization motif from the Kv2.1 potassium channel [101] to reduce neuropil fluorescence contamination thus improving single cell resolution, similarly to the strategy previously used to enable single-cell precision optogenetics [102,103]. Using a simple wide-field imaging set-up, soma-localized Voltron reported single action potentials in hippocampal parvalbumin neurons and visual cortex pyramidal neurons in awake mice. Particularly impressive was the ability to perform single spike detection in up to 46 GABAergic neurons in visual cortex layer I imaged simultaneously for over 15 minutes (Table 2). Voltron also retained its high performance in live zebrafish and fruit flies, reliably reporting single action potentials. Voltron relies on bright synthetic dyes delivered exogenously. While dye administration is straightforward, even for mice, easily passing through the blood brain barrier, quick internalization and accumulation of the JF dyes in the cytoplasm of the cells create high background fluorescence, significantly reducing imaging quality. Development of new dyes that fluoresce only upon binding to the HaloTag will significantly improve Voltron class of GEVIs.

Among all tested light-driven proton pumps, Arch uniquely possesses dim near-infrared fluorescence under strong red laser excitation [90]. Furthermore, Arch fluorescence intensifies as membrane potential increases. The mechanism of this voltage sensitive fluorescence is still poorly studied, but might involve chromophore protonation [92]. Hochbaum et al. subjected Arch to directed molecular evolution to improve its brightness and voltage sensitivity [59]. The result, dubbed QuasAr2, is capable of reporting single action potentials with high temporal resolution in intact brain tissue and even in mouse nodose ganglia in vivo [99] (Table 2). Near-infrared fluorescence and high voltage sensitivity of QuasAr2 enabled the recording of optogenetically-triggered synaptic inputs in single trials to probe synaptic strength in intact brain tissue [27]. However, due to its extremely low brightness, requiring up to three orders of magnitude higher light intensity than for GFP imaging, QuasAr2 did not become a widely used tool. Subsequent enhancement of QuasAr2 brightness generated several variants with distinct biochemical and photophysical characteristics (Figure 2c). For example, Adam Cohen's group developed a blue light activated version of QuasAr, named NovArch [104] and paQuasAr3 [60]. Near-infrared fluorescence of NovArch and paQuasAr3 can be reversibly enhanced by blue illumination without affecting voltage sensitivity (Table 2). This blue light enhanced brightness of paQuasAr3 is sufficient to perform population voltage imaging with single cell resolution in hippocampal cells of awake mice. At the same time, it complicates combining voltage imaging with optogenetic control by channelrhodopsins. In another study, significant improvement of near-infrared brightness was achieved by implementation of a robotic multidimensional directed evolution approach [65]. Large libraries, up to tens of millions of independent clones, expressed in mammalian cells were screened for improved brightness and localization using robotic cell picker. Selected variant Archon1 exhibited several fold higher fluorescence and nearly ideal membrane localization in neurons when expressed in vivo in mice, zebrafish, and worms. The soma-localized version of Archon1 enabled population voltage imaging in striatum, cortex, and hippocampus of behaving mice under a cheap and simple wide-field imaging set up (personal communication). In addition, Archon1 was the first reported GEVI to report single action potentials in zebrafish larvae. Despite several generations of improved Arch-derived voltage sensors, the sensors still suffer from very low fluorescence brightness in comparison to other classes of GEVIs.

In the past few years the field of GEVIs was strongly dominated by opsin-based sensors, however, opsin-based sensors all share one disappointing limitation: they do not exhibit voltage sensitivity under two-photon excitation. Although fundamentally and technically more challenging than wide-field imaging, two-photon microscopy offers increased imaging depth with higher spatial resolution in scattering tissue. For example, two-photon microscopy has been used for subcellular voltage imaging in live fruit flies using GEVIs [105–107] as well as dendritic [108] and somatic [109] voltage using chemical dyes. The development of the improved version of the ASAP sensor [110], named ASAP3, enabled two-photon voltage imaging in awake mice using a fully genetically-encoded probe [81]. The ASAP family of GEVIs is based on the insertion of a circularly permuted EGFP into the voltage-sensing domain from *Gallus gallus* of voltage-sensitive phosphatases between the third and fourth transmembrane domains (Figure 2d). ASAP3 development involved random mutagenesis of the linkers between circularly permuted EGFP and voltage-sensing domain to optimize coupling of four transmembrane helix movement during depolarization to chromophore protonation. Using the soma-localized version of ASAP3, Chavarha et al. performed reliable voltage imaging with single cell single spike resolution in visual cortex at a depth up to ~130 µm. Improved performance of ASAP3 compared to its precursors can be partially attributed to its kinetics. ASAP3 has a submillisecond on rate, but several-fold slower off kinetics, unlike opsin-based sensors (Table 2). This combination of kinetic parameters results in significant boarding of optical action potential waveforms thus increasing the effective number of photons emitted during spikes. However, fast two-photon imaging is still limited to a small imaging area and recording of more than one cell at a time still represent significant technical challenge. Additionally, wide-field imaging of opsin-based sensor is more cost efficient.

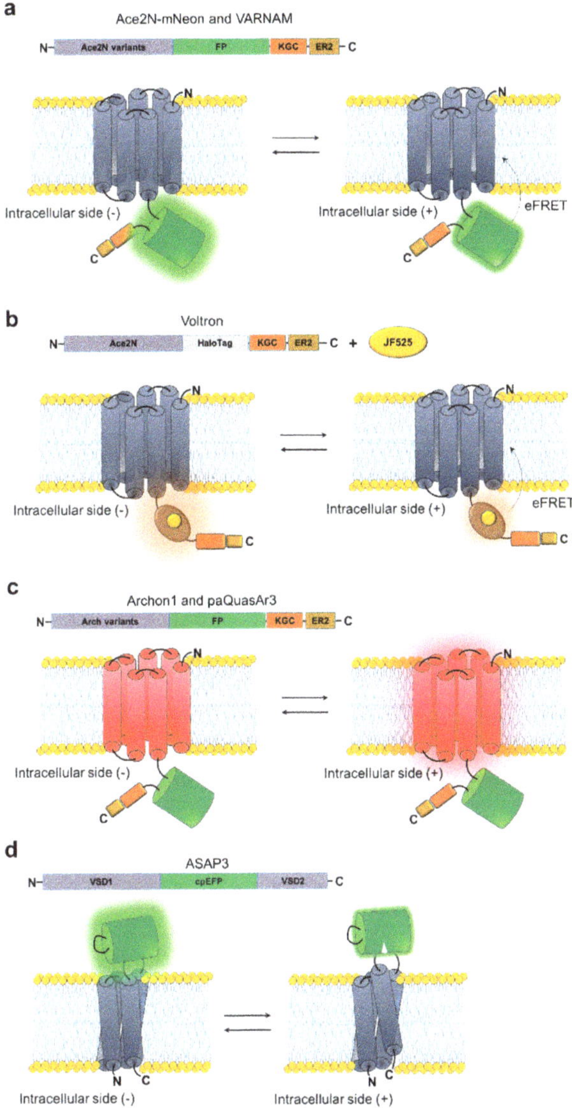

Figure 2. Molecular design of selected genetically-encoded and hybrid fluorescent voltage indicators. (**a**) eFRET-opsin voltage sensors Ace2N and VARNAM. (**b**) Hybrid eFRET-opsin voltage sensor Voltron. (**c**) Opsin-based voltage sensors Archon1 and paQuasAr3. (**d**) VSD-based voltage sensor ASAP3.

5. Neurotransmitter Indicators

Neurons communicate with one another using chemical messengers called neurotransmitters. Neurotransmitters are released by a pre-synaptic cell via vesicle-mediated exocytosis into the synaptic cleft [111], where they reach concentrations in the high micromolar range [112]. Here, neurotransmitters bind to receptors on the post-synaptic cell, initiating second messenger signaling cascades (via metabotropic receptors) or evoking ion flux (via ionotropic receptors) [113], and are subsequently cleared on the millisecond time scale from the post-synaptic cleft by neurotransmitter transporters [114]. In addition to classical neurotransmitters, including glutamate [115], gamma-aminobutyric acid (GABA) [116], and other biogenic amines [117], neurons also signal using non-classical neurotransmitters, such as peptides [118], endocannabinoids [119], and even gases [120]. Recent advance in protein engineering have yielded a number of fully genetically-encoded fluorescent sensors enabling real time optical detection of a few key neurotransmitters with high specificity and sensitivity within the intact brains of several model organisms.

All genetically-encoded neurotransmitter indicators (GENIs) utilize two types of ligand-binding proteins as sensing moieties, either periplasmic-binding proteins (PBP) or G protein-coupled receptors (GPCRs; Figure 3). PBPs are bacterial non-enzymatic receptors with specificity for a wide variety of small molecule ligands, including neurotransmitters. Upon ligand binding, PBPs undergo significant conformational changes [121] sufficient to modulate fluorescence of covalently bound optical probes thus providing versatile scaffolds for fluorescence biosensor engineering [122–124]. To date, PBPs have been used to develop three different classes of genetically-encoded neurotransmitter indicators (GENIs) specialized for the optical detection of glutamate, GABA, and glycine. The first GENIs, FLIPE and superGluSnFR, capable of reporting transients of the chief excitatory neurotransmitter glutamate in cultured neurons, were engineered by fusing the glutamate PBP GltI from *Escherichia coli* (also known as ybeJ) with a FRET pair of cyan fluorescent protein (CFP) and yellow fluorescent protein (YFP) [125,126]. However, a relatively low signal-to-noise ratio, an inherent drawback for most of FRET-based sensors, significantly limited applications of FLIPE and superGluSnFR in vivo. A breakthrough in imaging glutamate transients in vivo was achieved with the development of a single FP-based sensor, iGluSnFR [127], boasting a positive intensiometric fluorescence response (Table 3). iGluSnFR was constructed by insertion of a circularly permuted (cp) EGFP into the interdomain hinge region of GltI with further optimization of linkers between cpEGFP and GltI to improve the fluorescence dynamic range (Figure 3a). For expression in neurons, iGluSnFR is targeted to the extracellular side of the plasma membrane by fusion with a PDGFR peptide segment, similar to the first generation of glutamate sensors [125,126]. Two-photon microscopy of iGluSnFR allowed single trial recordings of glutamate transients at single dendritic spines as well as dendritic branches in the primary motor cortex of awake behaving mice [127]. In addition, iGluSnFR reliably reported synaptic activity in vivo in zebrafish and the nematode *Caenorhabditis elegans* [127]. However, due to its slow off rate, with a reported decay half-time of 92 ms, iGluSnFR failed to resolve fast transients associated with local glutamate release during external electrical stimulations of neurons higher than 10 Hz [128,129], highlighting a clear need for a glutamate sensor with improved biophysical characteristics.

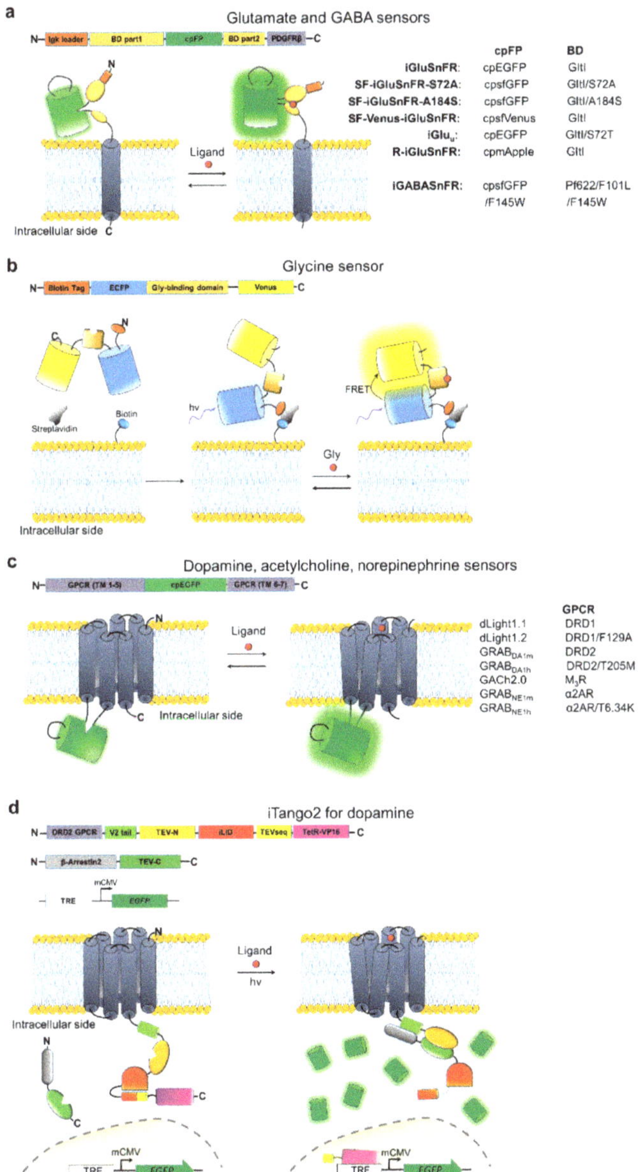

Figure 3. Molecular design of genetically-encoded neurotransmitter indicators. (**a**) PBP-based glutamate and GABA sensors. (**b**) PBP-based glycine sensor. (**c**) GPCR-based dopamine, acetylcholine, and norepinephtine sensors. (**d**) iTango2 system for dopamine detection.

Table 3. Characteristics of selected genetically-encoded fluorescent neurotransmitter reporters.

Sensor	Ligand	Ex/Em (nm)	ΔF/F (%)	K_d (μM)	On Kinetics (ms)	Off Kinetics (ms)	In Vivo Imaging	Ref
iGluSnFR	Glu [a]	490/510	100	4.9	15	92	Single dendritic spines	[127]
SF-iGluSnFR-A184S	Glu [a]	490/510	69	0.6	85	450	Single dendritic spines	[130]
SF-iGluSnFR-S72A	Glu [a]	490/510	250	34	5	11	Single dendritic spines	[130]
SF-Venus-iGluSnFR	Glu [a]	515/528	66	2.0	ND	ND	Single dendritic spines	[130]
$iGlu_u$	Glu [a]	490/510	170	53	0.7	3	Not tested	[129]
R-iGluSnFR1	Glu [a]	562/588	−33	18	ND	ND	Not tested	[131]
iGABASnFR	GABA	485/510	10 [b]	~9 [b]	~25 [c]	~60 [c]	Aggregated neuropil imaging	[132]
GlyFS	Gly	434/477 515/528 (FRET)	28.3 19.0	28 21	ND	ND	Not tested	[133]
dLight1.1	DA	490/516	230 [d]	0.33 [d]	ND	ND	Aggregated signal	[134]
dLight1.2	DA	490/516	340 [d]	0.77 [d]	9.5 [e]	90 [e]	Aggregated signal	[134]
$GRAB_{DA1m}$	DA	490/510	90	0.13	80 [e]	3100 [e]	Aggregated signal	[135]
$GRAB_{DA1h}$	DA	490/510	90	0.01	110 [e]	17,150 [e]	Aggregated signal	[135]
GACh2.0	ACh	490/510	90	2	280 [d]	760 [d]	Single neurons with 15 trails averaging	[136]
$GRAB_{NE1m}$	NE	490/510	230	1.9	72 [d]	680 [d]	Aggregated signal	[137]
$GRAB_{NE1h}$	NE	490/510	150	0.093	36 [d]	1890 [d]	Not tested	[137]
iTango2	DA	488/507 554/581	900	ND	~10^5-10^6	NA	Single neurons	[138]

Ex—excitation wavelength; Em—emission wavelength; ΔF/F—maximally achievable fluorescence changes between ligand-free and ligand saturated states; K_d—ligand dissociation constant measured for cultured neurons expressing corresponding sensor, unless otherwise stated; ND—not determined; NA—not applicable. [a] shows similar affinity for Asp; [b] characterized in vitro; [c] estimated from the trace for one action potential in cultured neurons presented in the corresponding publication; [d] measured in HEK cells; [e] measured in acute brain slice.

Further optimization of iGluSnFR by rational design resulted in variants with improved brightness, kinetics, and altered glutamate affinity. The substitution of cpEGFP with a circularly permutated version of superfolder GFP (cpsfGFP) significantly enhanced brightness and photostability giving rise to a new series of the SF-iGluSnFR glutamate sensors [130] (SF stands for superfolder GFP used as reporting moiety; Figure 3a). Subsequent mutagenesis of the GltI domain in SF-iGluSnFR generated two variants, named SF-iGluSnFR-A184S and SF-iGluSnFR-S72A, which exhibit higher affinity and faster kinetics, respectively, compared to the original iGluSnFR [130] (Table 3). The SF-iGluSnFR-A184S variant significantly outperformed iGluSnFR in detecting stimulus-evoked glutamate release at individual dendritic spines imaged in neuropils of the visual cortex under two-photon excitation. In turn, the faster off rate of SF-iGluSnFR-S72A enabled almost an order of magnitude higher temporal resolution than was previously achievable with original iGluSnFR. For example, SF-iGluSnFR-S72A fluorescence was able to resolve synaptic responses in brain tissue during single action potentials evoked at frequencies up to 100 Hz. Even higher temporal resolution was achieved by the iGluSnFR-S72T point mutant, named iGluu (u stands for "ultrafast") [129], which was able to resolve responses at single synapses in hippocampal slices during single action potentials evoked even at frequencies of 100 Hz. However, it showed about 1.5-fold lower sensitivity than SF-iGluSnFR-S72A under similar imaging conditions (Table 3). Overall, the kinetics and sensitivity of SF-iGluSnFRs and iGluu make them superior to the fastest green calcium sensor GCaMP6f in terms of monitoring presynaptic activity, as was demonstrated for granule cell axons within mouse cerebellar brain slices.

Mutagenesis of the fluorescence moiety of iGluSnFRs enabled diversification of the spectral properties of the glutamate sensor, highlighting the versatility of its molecular design. Introduction of point mutations in the chromophore forming tripeptide led to SF-iGluSnFR variants with blue (emission peak ~450 nm), cyan (emission peak ~475 nm), and yellow (emission peak ~530 nm) fluorescence [130]. On the other hand, substitution of cpEGFP with cpmApple from the R-GECO1 calcium sensor resulted in the R-iGluSnFR glutamate sensor with red emission peaked at ~590 nm [131]. Despite subsequent intense optimization by random mutagenesis resulting in an additional 12 amino acid mutations throughout the protein, performance of R-iGluSnFR is far inferior to that of green glutamate sensors, with a dynamic range of only -33% compared to 250% for SF-iGluSnFR-S72A, thus precluding its in vivo application. Among chromatic variants of iGluSnFR, only the sensor with yellow emission, SF-Venus-iGluSnFR, was utilized for in vivo imaging in mouse visual cortex. The major advantage of SF-Venus-iGluSnFR over green fluorescence sensors for glutamate is its high brightness under two-photon excitation at 1,030 nm, compatible with powerful femtosecond fiber lasers, which are gaining popularity in the neuroscience community due to their affordability and usability. Currently, fluorescent sensors for glutamate represent the most diverse group of GENIs, however, the most practical variants are still limited to green fluorescence.

The versatility of the PBP-based sensor design was also demonstrated by the development of green fluorescent sensors for the chief inhibitory neurotransmitter GABA, called an intensity-based GABA Sensing Fluorescence Reporter or iGABASnFR [132] (Figure 3a). Marvin et al. used the newly identified PBP Pf622 cloned from *Pseudomonas fluorescens* [139] as a GABA-binding domain. The initial version of the sensor, created by insertion of cpsfGFP into the Pf622 domain, was further optimized by side-directed mutagenesis to increase affinity and sensitivity (Table 3). The iGABASnFR was shown to report GABA release events over large volume of brain tissue in vivo in mouse visual cortex and in zebrafish cerebellum with sub-second kinetics, although with rather low fluorescence changes at just 10%. While iGABASnFR is a solely available fully genetically-encoded sensor for GABA, it will most likely require further optimization to enhance affinity, kinetics, and sensitivity before it can be widely adopted by the neuroscience community.

Development of the PBP-based sensor for Gly appeared to be more challenging. The wild-type PBP Atu2422 from *Agrobacterium tumefaciens*, selected as a binding domain for glycine sensor development, also displays promiscuous binding for L-serine and GABA [140]. To increase specificity of Atu2422, Zhang et al. employed computationally-guided mutagenesis. Introducing steric obstructions into the ligand-binding site of Atu2422 produced a binding domain highly selective for glycine. Fusing the engineered binding domain with CFP-YFP FRET pair, but not with cpEGFP as in iGluSnFR, resulted in the functional glycine sensor, denoted GlyFS, characterized by ~28% of fluorescence dynamic range (Table 3). Membrane targeting of GlyFS via a PDGFR peptide fusion, as used for the iGluSnFR and iGABASnFR sensors, reduced dynamic range to 4%, which is 7-fold lower than that in vitro, due to inefficient membrane trafficking of the protein. Instead, extracellular space immobilization was performed through a biotinylation-based technique that anchors proteins to the outer side of the plasma membrane (Figure 3b). Taking advantage of the GlyFS ratiometric response, the authors performed measurements of the absolute concentration of Gly in brain slices revealing dependency of Gly concentration on animal age. The suggested immobilization technique requires biotinylation of cell surface in intact brain tissue with subsequent delivery of the GlyFS sensor and streptavidin [141]. While the biotinylation-based technique was demonstrated to be compatible with tumor imaging in live mice [142], it might not be feasible for neuroimaging in vivo thus limiting GlyFS application to in vitro and ex vivo preparations. The GlyFS engineering is an inspiring example showing that PBPs are flexible templates for generating diversity of binding domains. Nevertheless, GlyFS still requires further development to increase its usefulness and applicability for neuroscience research.

The second class of ligand-binding domains utilized for GENI engineering is represented by GPCRs. GPCRs share a highly conservative 3D structure comprising of seven α-helix transmembrane domains (TM) and constitute the majority of receptors with high specificity for endogenous neurotransmitters. Structural studies indicate that the largest conformation changes upon ligand binding to GPCRs by occurs at the cytosolic end of TM6. The α-helix of TM6 can undergo more than 6 Å shift outwards from the helix bundle center at the same time initiating the 2–3 Å shift of TM5 [143,144]. For this reason, the third intracellular loop, located between TM5 and TM6, is a primary target for insertion of reporting moieties to engineer GENIs. Indeed, first generation of GPCR-based sensors was constructed by insertion of FRET donors and acceptors into the third intracellular loop and at the shortened C terminal domain [145–148]. The FRET-based sensors exhibited good specificity and affinity to agonists as well as adequate kinetics, however, their utility for in vivo imaging is limited due to the relatively modest ~10%, changes in FRET signal. Very recently, an alternative design of GENIs was suggested to overcome this limitation. Incorporation of the conformationally sensitive cpEGFP into the third intracellular loop enabled over an order of magnitude larger fluorescence changes during GPCR activation than for the FRET-based sensors while preserving high specificity and reasonable kinetics of fluorescence response (Figure 3c). Typically, this engineering strategy involves three consecutive steps: (i) optimization of the insertion site within third intracellular loop of GPCR; (ii) optimization of the linkers between cpEGFP and GPCR; (iii) tuning ligand affinity by point mutations within the ligand-binding pocket. Moreover, the bulky cpEGFP moiety introduced into the intracellular loop almost completely demolish coupling with major GPCR downstream pathways, presumably due to the steric hindrance imposed for G protein or arrestin interaction with the GPCR [149,150]. As a result, this design strategy was successfully exploited to develop a series of GPCR-based sensors suitable for in vivo imaging of dopamine (DA), acetylcholine (ACh), and norepinephrine (NE), also called noradrenaline (NA), in multiple model organisms. In addition, as a proof of concept a class of sensors for other neuromodulators and neuropeptides were engineered using G_s-coupled β1 and β2 adrenergic receptors, G_i-coupled κ and μ-type opioid receptors, and G_q-coupled serotonin receptor-2A and melatonin type-2 receptor [134].

Two groups in parallel independently developed green fluorescence DA sensors with similar biochemical characteristics (Table 3). Patriarchi et al. reported a series of DA sensors, called dLight, developed based on the human Dopamine Receptor D1 (DRD1) [134]. Two variants in this series, dLight1.1 and dLight1.2, showed optimal combination of apparent DA affinity and sensitivity and were used for in vivo imaging in behaving mice (Table 3). The higher affinity dLight1.1 sensor enabled visualization of spontaneous and optogenetically evoked dopamine transients in the nucleus accumbens by recording aggregated fluorescence signal from multiple cells using fiber photometry. Under these imaging conditions, dLight1.1 could report both activation and inhibition of dopamine transients resolving individual spikes at up to 5 Hz stimulation frequency. Higher spatial resolution of dopamine imaging was achieved under two-photon microscopy of dLight1.2 expressed in layer 2/3 neurons of the cortex during a visuomotor learning task. Two-photon imaging with micron-resolution revealed a dopamine transient map with functionally heterogeneous dopamine signals in the cortex. In contrast, utilizing Dopamine Receptor D2 (DRD2) as a sensing moiety, Sun et al. developed a pair of alternative dopamine sensors, named genetically-encoded GPCR-activation-based-DA sensors ($GRAB_{DA}$) [135]. Overall, in comparison to the dLight sensors, two selected variants, $GRAB_{DA1m}$ and $GRAB_{DA1m}$, are characterized by several fold smaller dynamic range and slower kinetics but higher apparent affinity to DA (Table 3). Similar to the dLight sensors, the signal-to-noise ratio and temporal resolution of $GRAB_{DA}$ were sufficient to record the dynamic bi-directional changes during DA activation and inhibition in dorsal striatum and nucleus accumbens in behaving mice using fiber photometry. In particular, imaging $GRAB_{DA}$ in nucleus accumbens revealed a time-locked DA elevation aligned to various sexual behaviors, confirming an important role of DA in behavioral motivation, anticipation, or arousal. In addition, the $GRAB_{DA}$ sensors were shown to readily respond to visually induced DA release in the intact brain of the zebrafish larvae as well as reveal compartmentalized DA dynamics in the mushroom body of fruit flies with single neuron resolution. In summary, two groups of the currently available DA sensors demonstrated comparable performance in vivo and selection of a particular variant have to be determined based on the exact intended application.

Similar design approaches enabled the development a pair of NE sensors with altered ligand affinity based on the α-adrenergic receptor (α2AR) [137]. Two NE sensors, named $GRAB_{NE1m}$ and $GRAB_{NE1h}$, are characterized by micromolar and nanomolar apparent affinity for NE, respectively, and exhibit several fold increase in fluorescence upon NE binding when measured in cultured cells (Table 3).

Due to its slightly higher dynamic range, $GRAB_{NE1m}$ is the preferred variant for in vivo imaging. For example, $GRAB_{NE1m}$ allowed reliable detection of NE release in locus coeruleus and hypothalamus in freely moving mice using fiber photometry. In addition, confocal imaging of $GRAB_{NE1m}$-expressing neurites in the optic tectum of zebrafish revealed a time-locked increase in green fluorescence during looming stimuli.

Utilizing the human muscarinic acetylcholine receptor 3 as a sensing moiety enabled development of green fluorescence sensor for ACh, named GACh2.0 [136]. While GACh2.0 exhibited similar dynamic range to that of the $GRAB_{DA}$ sensors and adequate affinity in cultured cells (Table 3), its in vivo performance was not sufficient for single-trial imaging of the ACh transients when tested under two-photon microscopy in visual cortex in mice. Nevertheless, in intact brain tissue GACh2.0 showed ~5% fluorescence changes in a single trial in medial entorhinal cortex during multiple pulses of electrical stimulation, though single pulse stimulation still could not be detected. The poor in vivo performance of GACh2.0 can be attributed to severe sensor mislocalization when expressed in brain tissue; further improvement will certainly require optimization of membrane trafficking in vivo. Furthermore, GACh2.0 activation exhibited downstream coupling to G_q-dependent calcium signaling. While the coupling was about seven-fold smaller compared to wild-type M_3R, it still was much higher than for other analogous GPCR-based neurotransmitter sensors. Therefore, the Sindbis viral expression system, which permitted a more rapid (~18 h) and robust expression, was used for the GACh2.0 expression in brain slices and in vivo, while other GPCR-based GENIs were shown to be performing well with the widely used adeno associated virus (AAV) gene delivery.

Activation of the native downstream-coupled pathways by GPCR-based GENIs is usually an undesired drawback due to potential artifacts. However, it is possible to engineer GPCR-coupled downstream pathways for optical signal activation, for example, by inducing the expression of a reporter gene [151–153]. One such fully genetically-encoded system is the Tango assay. The Tango system couples a transcription factor to a GPCR receptor via a specific tobacco etch virus (TEV) protease–sensitive cleavage site. The GPCR activation recruits TEV protease fused to β-arrestin, which releases tethered transcription factor (tTA) by cleavage TEV site, initiating expression of a reporter gene [152]. Signal amplification via gene expression enables single-cell resolution detection of neurotransmitter release with nanomolar sensitivity. Recent modification of the Tango system, named DRD2-iTango2, enabled its application for in vivo detection of DA release in the mouse nucleus accumbens during locomotion. DA release was read out by GFP expression activated via dopamine receptor 2. As a main difference to the first generation Tango systems, iTango2 incorporated improved light-inducible dimer [154] (iLID; Figure 3d). This modification prevented access by TEV protease in the dark state thus significantly improving signal-to-noise ratio and reducing background fluorescence, thus overcoming the main caveats of the original Tango system. As a result, iTango2 represents three-component fully genetically-encoded system with highly modular design (Figure 3d), that can be easily adopted to the end user needs. For example, the iTango2 system was demonstrated to work well in cultured cells with various GPCRs, such as neuropeptide Y receptor type 1, cannabinoid receptor type 1, and serotonin receptor 1A. At the same time, iTango2 can be designed to conditionally express any gene of interest. For instance, using a light-driven chloride pump as a downstream expressing gene it was possible to gain optogenetic control over the behavioral relevance of a temporally and genetically identified population of neurons. In comparison to the most of the GENIs described above, the iTango2 system enables read out with single cell resolution, rather than recording aggregated signal from multiple cells expressing sensor (Table 3). However, higher spatial resolution comes at the cost of significantly reduced temporal resolution, on the order of tens of minutes, as well as the inability to record inhibition of neurotransmitter release. In addition, DRD2-iTango requires efficient co-expression of three genes packed into the separate AAVs that may impose optimization of multiple AAV co-expression for different cells types and brain regions.

6. Conclusions

Genetically-encoded sensors are indispensable tools for neuroscientists because they enable the study of neural population dynamics, circuit organization, and the activity patterns of specific cell types and even distinct organelles. We predict sensors for key signaling molecules will continue to improve in terms of brightness, kinetics, localization, toxicity, and red-shifted variants, expanding the toolbox for neuroscientists in important and exciting ways. For example, expanded genome databases [155,156] and directed evolution strategies [157,158] expand the resources available to protein engineers to identify and optimize ligand binding domains.

The development of new imaging modalities, such as prisms to image across midline fissures [159,160] and gradient index (GRIN) lenses to image deep brain structures [161,162], are beginning to afford optical access to historically inaccessible brain regions, and, in combination with the development of new sensors, facilitate more vigorous interrogation of brain function. Optical indicators are also amenable to methods aimed at manipulating brain activity, such as optogenetic and chemogenetic actuators, permitting the simultaneous observation and manipulation of neural activity. Parallel advances in complementary technologies, such as brain clearing and tissue expansion [163,164], further broaden the power and applicability of sensors to map distinct circuits [165]. Future studies monitoring neural activity in vivo will continue to benefit from efforts to improve genetically-encoded fluorescent sensors.

Author Contributions: Conceptualization, K.D.P.; writing—original draft preparation, K.D.P., F.V.S.; writing—review and editing, K.D.P., M.H.M.; visualization, K.D.P., F.V.S.; supervision, K.D.P.; funding acquisition, F.V.S.

Funding: This research was funded by Russian Science Foundation grant No. 16-15-10323 and Russian Foundation for Basic Research No. 19-04-00395.

Conflicts of Interest: The authors declare no conflict of interest.

References

1. Deisseroth, K.; Feng, G.; Majewska, A.K.; Miesenbock, G.; Ting, A.; Schnitzer, M.J. Next-Generation Optical Technologies for Illuminating Genetically Targeted Brain Circuits. *J. Neurosci.* **2006**, *26*, 10380–10386. [CrossRef] [PubMed]
2. Knopfel, T.; Lin, M.Z.; Levskaya, A.; Tian, L.; Lin, J.Y.; Boyden, E.S. Toward the Second Generation of Optogenetic Tools. *J. Neurosci.* **2010**, *30*, 14998–15004. [CrossRef] [PubMed]
3. Kügler, S.; Kilic, E.; Bähr, M. Human synapsin 1 gene promoter confers highly neuron-specific long-term transgene expression from an adenoviral vector in the adult rat brain depending on the transduced area. *Gene Ther.* **2003**, *10*, 337–347. [CrossRef] [PubMed]
4. Sauer, B. Inducible gene targeting in mice using the Cre/lox system. *Methods* **1998**, *14*, 381–392. [CrossRef] [PubMed]
5. Madisen, L.; Garner, A.R.; Shimaoka, D.; Chuong, A.S.; Klapoetke, N.C.; Li, L.; Van der Bourg, A.; Niino, Y.; Egolf, L.; Monetti, C. Transgenic mice for intersectional targeting of neural sensors and effectors with high specificity and performance. *Neuron* **2015**, *85*, 942–958. [CrossRef] [PubMed]
6. Daigle, T.L.; Madisen, L.; Hage, T.A.; Valley, M.T.; Knoblich, U.; Larsen, R.S.; Takeno, M.M.; Huang, L.; Gu, H.; Larsen, R. A Suite of Transgenic Driver and Reporter Mouse Lines with Enhanced Brain-Cell-Type Targeting and Functionality. *Cell* **2018**, *174*, 465–480. [CrossRef]
7. Margolis, D.J.; Lütcke, H.; Schulz, K.; Haiss, F.; Weber, B.; Kügler, S.; Hasan, M.T.; Helmchen, F. Reorganization of cortical population activity imaged throughout long-term sensory deprivation. *Nat. Neurosci.* **2012**, *15*, 1539–1546. [CrossRef]
8. Huber, D.; Gutnisky, D.A.; Peron, S.; O'Connor, D.H.; Wiegert, J.S.; Tian, L.; Oertner, T.G.; Looger, L.L.; Svoboda, K. Multiple dynamic representations in the motor cortex during sensorimotor learning. *Nature* **2012**, *484*, 473–478. [CrossRef]
9. Dana, H.; Mohar, B.; Sun, Y.; Narayan, S.; Gordus, A.; Hasseman, J.P.; Tsegaye, G.; Holt, G.T.; Hu, A.; Walpita, D. Sensitive red protein calcium indicators for imaging neural activity. *Elife* **2016**, *5*, 1–24. [CrossRef]

10. Zucker, R.S. Calcium- and activity-dependent synaptic plasticity. *Curr. Opin. Neurobiol.* **1999**, *9*, 305–313. [CrossRef]
11. Grienberger, C.; Konnerth, A. Imaging Calcium in Neurons. *Neuron* **2012**, *73*, 862–885. [CrossRef] [PubMed]
12. Berridge, M.J.; Lipp, P.; Bootman, M.D. The versatility and universality of calcium signalling. *Nat. Rev. Mol. Cell Biol.* **2000**, *1*, 11–21. [CrossRef]
13. Helmchen, F.; Borst, J.G.; Sakmann, B. Calcium dynamics associated with a single action potential in a CNS presynaptic terminal. *Biophys. J.* **1997**, *72*, 1458–1471. [CrossRef]
14. Koester, H.J.; Sakmann, B. Calcium dynamics associated with action potentials in single nerve terminals of pyramidal cells in layer 2/3 of the young rat neocortex. *J. Physiol.* **2000**, *529*, 625–646. [CrossRef] [PubMed]
15. Broussard, G.J.; Liang, R.; Tian, L. Monitoring activity in neural circuits with genetically encoded indicators. *Front. Mol. Neurosci.* **2014**, *7*, 97. [CrossRef] [PubMed]
16. Lin, M.Z.; Schnitzer, M.J. Genetically encoded indicators of neuronal activity. *Nat. Neurosci.* **2016**, *19*, 1142–1153. [CrossRef] [PubMed]
17. Miyawaki, A.; Llopis, J.; Heim, R.; Michael McCaffery, J.; Adams, J.A.; Ikura, M.; Tsien, R.Y. Fluorescent indicators for Ca^{2+} based on green fluorescent proteins and calmodulin. *Nature* **1997**. [CrossRef]
18. Baird, G.S.; Zacharias, D.A.; Tsien, R.Y. Circular permutation and receptor insertion within green fluorescent proteins. *Proc. Natl. Acad. Sci. USA* **1999**. [CrossRef]
19. Nagai, T.; Sawano, A.; Park, E.S.; Miyawaki, A. Circularly permuted green fluorescent proteins engineered to sense Ca2+. *Proc. Natl. Acad. Sci. USA* **2001**. [CrossRef]
20. Nakai, J.; Ohkura, M.; Imoto, K. A high signal-to-noise Ca^{2+} probe composed of a single green fluorescent protein. *Nat. Biotechnol.* **2001**. [CrossRef]
21. Pérez Koldenkova, V.; Nagai, T. Genetically encoded Ca^{2+} indicators: Properties and evaluation. *Biochim. Biophys. Acta Mol. Cell Res.* **2013**, *1833*, 1787–1797. [CrossRef] [PubMed]
22. Greenwald, E.C.; Mehta, S.; Zhang, J. Genetically Encoded Fluorescent Biosensors Illuminate the Spatiotemporal Regulation of Signaling Networks. *Chem. Rev.* **2018**, *118*, 11707–11794. [CrossRef] [PubMed]
23. Ziv, Y.; Burns, L.D.; Cocker, E.D.; Hamel, E.O.; Ghosh, K.K.; Kitch, L.J.; El Gamal, A.; Schnitzer, M.J. Long-term dynamics of CA1 hippocampal place codes. *Nat. Neurosci.* **2013**. [CrossRef] [PubMed]
24. Ahrens, M.B.; Orger, M.B.; Robson, D.N.; Li, J.M.; Keller, P.J. Whole-brain functional imaging at cellular resolution using light-sheet microscopy. *Nat. Methods* **2013**. [CrossRef] [PubMed]
25. Sheffield, M.E.J.; Dombeck, D.A. Calcium transient prevalence across the dendritic arbour predicts place field properties. *Nature* **2014**. [CrossRef] [PubMed]
26. Chen, T.W.; Wardill, T.J.; Sun, Y.; Pulver, S.R.; Renninger, S.L.; Baohan, A.; Schreiter, E.R.; Kerr, R.A.; Orger, M.B.; Jayaraman, V. Ultrasensitive fluorescent proteins for imaging neuronal activity. *Nature* **2013**. [CrossRef] [PubMed]
27. Fan, L.Z.; Nehme, R.; Adam, Y.; Jung, E.S.; Wu, H.; Eggan, K.; Arnold, D.B.; Cohen, A.E. All-optical synaptic electrophysiology probes mechanism of ketamine-induced disinhibition. *Nat. Methods* **2018**, *15*, 823–831. [CrossRef] [PubMed]
28. Broussard, G.J.; Liang, Y.; Fridman, M.; Unger, E.K.; Meng, G.; Xiao, X.; Na, J.; Leopoldo, P.; Tian, L. In vivo measurement of afferent activity with axon-specific calcium imaging. *Nat. Neurosci.* **2018**. [CrossRef]
29. Steinmetz, N.A.; Buetfering, C.; Lecoq, J.; Lee, C.R.; Peters, A.J.; Jacobs, E.A.K.; Coen, P.; Ollerenshaw, D.R.; Valley, M.T.; de Vries, S.E.J. Aberrant Cortical Activity in Multiple GCaMP6-Expressing Transgenic Mouse Lines. *Eneuro* **2017**, *4*, ENEURO.0207-17. [CrossRef]
30. McMahon, S.M.; Jackson, M.B. An Inconvenient Truth: Calcium Sensors Are Calcium Buffers. *Trends Neurosci.* **2018**. [CrossRef]
31. Yang, Y.; Liu, N.; He, Y.; Liu, Y.; Ge, L.; Zou, L.; Song, S.; Xiong, W.; Liu, X. Improved calcium sensor GCaMP-X overcomes the calcium channel perturbations induced by the calmodulin in GCaMP. *Nat. Commun.* **2018**, *9*, 1504. [CrossRef] [PubMed]
32. Tian, L.; Hires, S.A.; Mao, T.; Huber, D.; Chiappe, M.E.; Chalasani, S.H.; Petreanu, L.; Akerboom, J.; McKinney, S.A.; Schreiter, E.R. Imaging neural activity in worms, flies and mice with improved GCaMP calcium indicators. *Nat. Methods* **2009**. [CrossRef] [PubMed]
33. Barykina, N.V.; Subach, O.M.; Piatkevich, K.D.; Jung, E.E.; Malyshev, A.Y.; Smirnov, I.V.; Bogorodskiy, A.O.; Borshchevskiy, V.; Varizhuk, A.; Pozmogova, G.E. Green fluorescent genetically encoded calcium indicator based on calmodulin/M13-peptide from fungi. *PLoS ONE* **2017**, *12*, e0183757. [CrossRef] [PubMed]

34. Zhao, Y.; Araki, S.; Wu, J.; Teramoto, T.; Chang, Y.F.; Nakano, M.; Abdelfattah, A.S.; Fujiwara, M.; Ishihara, T.; Nagai, T. An Expanded Palette of Genetically Encoded Ca^{2+} Indicators -Supplementay Info. *Science* **2011**. [CrossRef] [PubMed]
35. Zhao, Y.; Abdelfattah, A.S.; Zhao, Y.; Ruangkittisakul, A.; Ballanyi, K.; Campbell, R.E.; Harrison, D.J. Microfluidic cell sorter-aided directed evolution of a protein-based calcium ion indicator with an inverted fluorescent response. *Integr. Biol.* **2014**. [CrossRef] [PubMed]
36. Sun, X.R.; Badura, A.; Pacheco, D.A.; Lynch, L.A.; Schneider, E.R.; Taylor, M.P.; Hogue, I.B.; Enquist, L.W.; Murthy, M.; Wang, S.S.H. Fast GCaMPs for improved tracking of neuronal activity. *Nat. Commun.* **2013**. [CrossRef] [PubMed]
37. Dana, H.; Sun, Y.; Mohar, B.; Hulse, B.; Hasseman, J.P.; Tsegaye, G.; Tsang, A.; Wong, A.; Patel, R.; Macklin, J.J. High-performance GFP-based calcium indicators for imaging activity in neuronal populations and microcompartments. *bioRxiv* **2018**. [CrossRef]
38. Barykina, N.V.; Subach, O.M.; Doronin, D.A.; Sotskov, V.P.; Roshchina, M.A.; Kunitsyna, T.A.; Malyshev, A.Y.; Smirnov, I.V.; Azieva, A.M.; Sokolov, I.S. A new design for a green calcium indicator with a smaller size and a reduced number of calcium-binding sites. *Sci. Rep.* **2016**, *6*, 34447. [CrossRef]
39. Barykina, N.V.; Doronin, D.A.; Subach, O.M.; Sotskov, V.P.; Plusnin, V.V.; Ivleva, O.A.; Gruzdeva, A.M.; Kunitsyna, T.A.; Ivashkina, O.I.; Lazutkin, A.A. NTnC-like genetically encoded calcium indicator with a positive and enhanced response and fast kinetics. *Sci. Rep.* **2018**. [CrossRef]
40. Moeyaert, B.; Holt, G.; Madangopal, R.; Perez-Alvarez, A.; Fearey, B.C.; Trojanowski, N.F.; Ledderose, J.; Zolnik, T.A.; Das, A.; Patel, D. Improved methods for marking active neuron populations. *Nat. Commun.* **2018**. [CrossRef]
41. Doronin, D.A.; Barykina, N.V.; Subach, O.M.; Sotskov, V.P.; Plusnin, V.V.; Ivleva, O.A.; Isaakova, E.A.; Varizhuk, A.M.; Pozmogova, G.E.; Malyshev, A.Y. Genetically encoded calcium indicator with NTnC-like design and enhanced fluorescence contrast and kinetics. *BMC Biotechnol.* **2018**, *18*, 10. [CrossRef] [PubMed]
42. Ohkura, M.; Sasaki, T.; Kobayashi, C.; Ikegaya, Y.; Nakai, J. An improved genetically encoded red fluorescent Ca^{2+} indicator for detecting optically evoked action potentials. *PLoS ONE* **2012**, *7*, e1796. [CrossRef] [PubMed]
43. Shen, Y.; Dana, H.; Abdelfattah, A.S.; Patel, R.; Shea, J.; Molina, R.S.; Rawal, B.; Rancic, V.; Chang, Y.F.; Wu, L. A genetically encoded Ca^{2+} indicator based on circularly permutated sea anemone red fluorescent protein eqFP578. *BMC Biol.* **2018**, *16*, 1–16. [CrossRef] [PubMed]
44. Lanin, A.A.; Chebotarev, A.S.; Barykina, N.V.; Subach, F.V.; Zheltikov, A.M. The whither of bacteriophytochrome-based near-infrared fluorescent proteins: Insights from two-photon absorption spectroscopy. *J. Biophotonics* **2019**, 1–7. [CrossRef] [PubMed]
45. Qian, Y.; Piatkevich, K.D.; Larney, B.; Abdelfattah, A.S.; Sohum, M.; Murdock, M.H.; Gottschalk, S.; Molina, R.S.; Zhang, W.; Chen, Y. A genetically encoded near-infrared fluorescent calcium ion indicator. *Nat. Methods* **2019**, *1*, 1–7. [CrossRef]
46. Qian, Y.; Rancic, V.; Wu, J.; Ballanyi, K.; Campbell, R.E. A bioluminescent Ca^{2+} indicator based on a topological variant of GCaMP6s. *ChemBioChem* **2018**, 1–6. [CrossRef]
47. Shen, Y.; Wu, S.-Y.; Rancic, V.; Aggarwal, A.; Qian, Y.; Miyashita, S.-I.; Ballanyi, K.; Campbell, R.E.; Dong, M. Genetically encoded fluorescent indicators for imaging intracellular potassium ion concentration. *Commun. Biol.* **2019**, *2*, 18. [CrossRef] [PubMed]
48. Bischof, H.; Rehberg, M.; Stryeck, S.; Artinger, K.; Eroglu, E.; Waldeck-Weiermair, M.; Gottschalk, B.; Rost, R.; Deak, A.T.; Niedrist, T. Novel genetically encoded fluorescent probes enable real-Time detection of potassium in vitro and in vivo. *Nat. Commun.* **2017**, *8*, 1–11. [CrossRef] [PubMed]
49. Görlach, A.; Klappa, P.; Kietzmann, D.T. The Endoplasmic Reticulum: Folding, Calcium Homeostasis, Signaling, and Redox Control. *Antioxid. Redox Signal.* **2006**. [CrossRef]
50. Marchi, S.; Patergnani, S.; Missiroli, S.; Morciano, G.; Rimessi, A.; Wieckowski, M.R.; Giorgi, C.; Pinton, P. Mitochondrial and endoplasmic reticulum calcium homeostasis and cell death. *Cell Calcium* **2018**. [CrossRef]
51. Giorgi, C.; Agnoletto, C.; Bononi, A.; Bonora, M.; de Marchi, E.; Marchi, S.; Missiroli, S.; Patergnani, S.; Poletti, F.; Rimessi, A. Mitochondrial calcium homeostasis as potential target for mitochondrial medicine. *Mitochondrion* **2012**. [CrossRef] [PubMed]
52. Otera, H.; Ishihara, N.; Mihara, K. New insights into the function and regulation of mitochondrial fission. *Biochim. Biophys. Acta Mol. Cell Res.* **2013**. [CrossRef] [PubMed]

53. Shui, B.; Wang, Q.; Lee, F.; Byrnes, L.J.; Chudakov, D.M.; Lukyanov, S.A.; Sonderman, H.; Kotlikoff, M.I. Circular permutation of red fluorescent proteins. *PLoS ONE* **2011**. [CrossRef] [PubMed]
54. Akerboom, J.; Carreras Calderón, N.; Tian, L.; Wabnig, S.; Prigge, M.; Tolö, J.; Gordus, A.; Orger, M.B.; Severi, K.E.; Macklin, J.J. Genetically encoded calcium indicators for multi-color neural activity imaging and combination with optogenetics. *Front. Mol. Neurosci.* **2013**, *6*, 1–29. [CrossRef]
55. Shen, Y.; Wiens, M.D.; Campbell, R.E. A photochromic and thermochromic fluorescent protein. *RSC Adv.* **2014**. [CrossRef]
56. Wu, J.; Liu, L.; Matsuda, T.; Zhao, Y.; Rebane, A.; Drobizhev, M.; Chang, Y.-F.; Araki, S.; Arai, Y.; March, K. Improved orange and red Ca^{2+} indicators and photophysical considerations for optogenetic applications. *ACS Chem. Neurosci.* **2013**. [CrossRef] [PubMed]
57. Wu, J.; Abdelfattah, A.S.; Miraucourt, L.S.; Kutsarova, E.; Ruangkittisakul, A.; Zhou, H.; Ballanyi, K.; Wicks, G.; Drobizhev, M.; Rebane, A. A long Stokes shift red fluorescent Ca^{2+} indicator protein for two-photon and ratiometric imaging. *Nat. Commun.* **2014**. [CrossRef] [PubMed]
58. Inoue, M.; Takeuchi, A.; Horigane, S.I.; Ohkura, M.; Gengyo-Ando, K.; Fujii, H.; Kamijo, S.; Takemoto-Kimura, S.; Kano, M.; Nakai, J. Rational design of a high-affinity, fast, red calcium indicator R-CaMP2. *Nat. Methods* **2014**. [CrossRef]
59. Hochbaum, D.R.; Zhao, Y.; Farhi, S.L.; Klapoetke, N.; Werley, C.A.; Kapoor, V.; Zou, P.; Kralj, J.M.; Maclaurin, D.; Smedemark-Margulies, N. All-optical electrophysiology in mammalian neurons using engineered microbial rhodopsins. *Nat. Methods* **2014**, *11*, 825–833. [CrossRef]
60. Adam, Y.; Kim, J.J.; Lou, S.; Cohen, A.E. All-optical electrophysiology reveals brain-state dependent changes in hippocampal subthreshold dynamics and excitability. *bioRxiv* **2018**. [CrossRef]
61. Klapoetke, N.C.; Murata, Y.; Kim, S.S.; Pulver, S.R.; Birdsey-Benson, A.; Cho, Y.K.; Morimoto, T.K.; Chuong, A.S.; Carpenter, E.J.; Tian, Z. Independent optical excitation of distinct neural populations. *Nat. Methods* **2014**, *11*, 338–346. [CrossRef] [PubMed]
62. Trojanowski, N.F.; Nelson, M.D.; Flavell, S.W.; Fang-Yen, C.; Raizen, D.M. Distinct Mechanisms Underlie Quiescence during Two Caenorhabditis elegans Sleep-Like States. *J. Neurosci.* **2015**, *35*, 14571–14584. [CrossRef] [PubMed]
63. Piatkevich, K.D.; Subach, F.V.; Verkhusha, V.V. Engineering of bacterial phytochromes for near-infrared imaging, sensing, and light-control in mammals. *Chem. Soc. Rev.* **2013**, *42*, 3441–3452. [CrossRef] [PubMed]
64. Piatkevich, K.D.; Suk, H.-J.; Kodandaramaiah, S.B.; Yoshida, F.; DeGennaro, E.M.; Drobizhev, M.; Hughes, T.E.; Desimone, R.; Boyden, E.S.; Verkhusha, V.V. Near-Infrared Fluorescent Proteins Engineered from Bacterial Phytochromes in Neuroimaging. *Biophys. J.* **2017**, *113*, 2299–2309. [CrossRef]
65. Piatkevich, K.D.; Jung, E.E.; Straub, C.; Linghu, C.; Park, D.; Suk, H.-J.; Hochbaum, D.R.; Goodwin, D.; Pnevmatikakis, E.; Pak, N. A robotic multidimensional directed evolution approach applied to fluorescent voltage reporters. *Nat. Chem. Biol.* **2018**. [CrossRef] [PubMed]
66. Yu, D.; Baird, M.A.; Allen, J.R.; Howe, E.S.; Klassen, M.P.; Reade, A.; Makhijani, K.; Song, Y.; Liu, S.; Murthy, Z. A naturally monomeric infrared fluorescent protein for protein labeling in vivo. *Nat. Methods* **2015**, *12*, 763–765. [CrossRef] [PubMed]
67. Filonov, G.S.; Piatkevich, K.D.; Ting, L.-M.; Zhang, J.; Kim, K.; Verkhusha, V.V. Bright and stable near-infrared fluorescent protein for in vivo imaging. *Nat. Biotechnol.* **2011**, *29*, 757–761. [CrossRef] [PubMed]
68. Piatkevich, K.D.; Subach, F.V.; Verkhusha, V.V. Far-red light photoactivatable near-infrared fluorescent proteins engineered from a bacterial phytochrome. *Nat. Commun.* **2013**, *4*, 2153. [CrossRef]
69. Rumyantsev, K.A.; Shcherbakova, D.M.; Zakharova, N.I.; Emelyanov, A.V.; Turoverov, K.K.; Verkhusha, V.V. Minimal domain of bacterial phytochrome required for chromophore binding and fluorescence. *Sci. Rep.* **2015**, *5*, 18348. [CrossRef]
70. Fosque, B.F.; Sun, Y.; Dana, H.; Yang, C.T.; Ohyama, T.; Tadross, M.R.; Patel, R.; Zlatic, M.; Kim, D.S.; Ahrens, M.B. Labeling of active neural circuits in vivo with designed calcium integrators. *Science* **2015**. [CrossRef]
71. Stringer, C.; Pachitariu, M.; Steinmetz, N.; Reddy, C.; Carandini, M.; Harris, K.D. Spontaneous behaviors drive multidimensional, brain-wide neural activity. *bioRxiv* **2018**. [CrossRef]
72. Saito, K.; Chang, Y.F.; Horikawa, K.; Hatsugai, N.; Higuchi, Y.; Hashida, M.; Yoshida, Y.; Matsuda, T.; Arai, Y.; Nagai, T. Luminescent proteins for high-speed single-cell and whole-body imaging. *Nat. Commun.* **2012**, *3*, 1262–1269. [CrossRef] [PubMed]

73. Rose, A.M.; Valdes, R. Understanding the sodium pump and its relevance to disease. *Clin. Chem.* **1994**, *40*, 1674–1685. [PubMed]
74. Rose, C.R.; Konnerth, A. NMDA receptor-mediated Na^+ signals in spines and dendrites. *J. Neurosci.* **2001**, *21*, 4207–4214. [CrossRef] [PubMed]
75. Rose, C.R. Book Review: Na^+ Signals at Central Synapses. *Neurosci.* **2002**. [CrossRef] [PubMed]
76. Llinás, R.R. The intrinsic electrophysiological properties of mammalian neurons: Insights into central nervous system function. *Science* **1988**, *242*, 1654–1664. [CrossRef] [PubMed]
77. Bean, B.P. The action potential in mammalian central neurons. *Nat. Rev. Neurosci.* **2007**, *8*, 451–465. [CrossRef]
78. Stuart, G.J.; Spruston, N. Dendritic integration: 60 years of progress. *Nat. Neurosci.* **2015**, *18*, 1713–1721. [CrossRef]
79. Peterka, D.S.; Takahashi, H.; Yuste, R. Imaging voltage in neurons. *Neuron* **2011**, *69*, 9–21. [CrossRef]
80. Kannan, M.; Vasan, G.; Huang, C.; Haziza, S.; Li, J.Z.; Inan, H.; Schnitzer, M.J.; Pieribone, V.A. Fast, in vivo voltage imaging using a red fluorescent indicator. *Nat. Methods* **2018**, *15*, 1108–1116. [CrossRef]
81. Chavarha, M.; Villette, V.; Dimov, I.K.; Pradhan, L.; Evans, S.W.; Shi, D.; Yang, R.; Chamberland, S.; Bradley, J.; Mathieu, B.; et al. Fast two-photon volumetric imaging of an improved voltage indicator reveals electrical activity in deeply located neurons in the awake brain. *bioRxiv* **2018**. [CrossRef]
82. Abdelfattah, A.S.; Kawashima, T.; Singh, A.; Novak, O.; Mensh, B.D.; Paninski, L.; Macklin, J.J.; Podgorski, K.; Lin, B.J.; Chen, T.W. Bright and photostable chemigenetic indicators for extended in vivo voltage imaging. *bioRxiv* **2018**. [CrossRef]
83. Siegel, M.S.; Isacoff, E.Y. A genetically encoded optical probe of membrane voltage. *Neuron* **1997**. [CrossRef]
84. Gong, Y.; Huang, C.; Li, J.Z.; Grewe, B.F.; Zhang, Y.; Eismann, S.; Schnitzer, M.J. High-speed recording of neural spikes in awake mice and flies with a fluorescent voltage sensor. *Science* **2015**, *350*, 1361–1366. [CrossRef] [PubMed]
85. Yizhar, O.; Fenno, L.; Zhang, F.; Hegemann, P.; Diesseroth, K. Microbial Opsins: A Family of Single-Component Tools for Optical Control of Neural Activity Topic Introduction Microbial Opsins: A Family of Single-Component Tools for Optical Control of Neural Activity. *Cold Spring Harb. Protoc.* **2011**, 273–283. [CrossRef]
86. Mattis, J.; Tye, K.M.; Ferenczi, E.A.; Ramakrishnan, C.; Shea, D.J.O.; Prakash, R.; Gunaydin, L.A.; Hyun, M.; Fenno, L.E.; Gradinaru, V.P. rinciples for applying optogenetic tools derived from direct comparative analysis of microbial opsins. *Nat. Methods* **2011**, *9*, 159–172. [CrossRef]
87. Nagel, G.; Szellas, T.; Huhn, W.; Kateriya, S.; Adeishvili, N.; Berthold, P.; Ollig, O.; Hegemann, P.; Bamberg, E. Channelrhodopsin-2, a directly light-gated cation-selective membrane channel. *Proc. Natl. Acad. Sci. USA* **2003**, *100*, 13940–13945. [CrossRef] [PubMed]
88. Gradinaru, V.; Thompson, K.R.; Deisseroth, K. eNpHR: A Natronomonas halorhodopsin enhanced for optogenetic applications. *Brain Cell Biol.* **2008**, *36*, 129–139. [CrossRef]
89. Kralj, J.M.; Hochbaum, D.R.; Douglass, A.D.; Cohen, A.E. Electrical spiking in Escherichia coli probed with a fluorescent voltage-indicating protein. *Science* **2011**. [CrossRef]
90. Kralj, J.M.; Douglass, A.D.; Hochbaum, D.R.; MacLaurin, D.; Cohen, A.E. Optical recording of action potentials in mammalian neurons using a microbial rhodopsin. *Nat. Methods* **2012**. [CrossRef]
91. Lanyi, J. Bacteriorhodopsin. *Annu. Rev. Physiol.* **2004**, *66*, 665–688. [CrossRef] [PubMed]
92. Maclaurin, D.; Venkatachalam, V.; Lee, H.; Cohen, A.E. Mechanism of voltage-sensitive fluorescence in a microbial rhodopsin. *Proc. Natl. Acad. Sci. USA* **2013**, *110*, 5939–5944. [CrossRef] [PubMed]
93. Kuznetsova, S.; Zauner, G.; Aartsma, T.J.; Engelkamp, H.; Hatzakis, N.; Rowan, A.E.; Nolte, R.J.M.; Christianen, P.C.M.; Canters, G.W. The enzyme mechanism of nitrite reductase studied at single-molecule level. *Proc. Natl. Acad. Sci. USA* **2008**, *105*, 3250–3255. [CrossRef] [PubMed]
94. Bayraktar, H.; Fields, A.P.; Kralj, J.M.; Spudich, J.L.; Rothschild, K.J.; Cohen, A.E. Ultrasensitive measurements of microbial rhodopsin photocycles using photochromic FRET. *Photochem. Photobiol.* **2012**, *88*, 90–97. [CrossRef] [PubMed]
95. Gong, Y.; Wagner, M.J.; Li, J.Z.; Schnitzer, M.J. Imaging neural spiking in brain tissue using FRET-opsin protein voltage sensors. *Nat. Commun.* **2014**, *5*, 1–11. [CrossRef] [PubMed]
96. Zou, P.; Zhao, Y.; Douglass, A.D.; Hochbaum, D.R.; Brinks, D.; Werley, C.A.; Harrison, D.J.; Campbell, R.E.; Cohen, A.E. Bright and fast multicoloured voltage reporters via electrochromic FRET. *Nat. Commun.* **2014**, *5*, 4625. [CrossRef] [PubMed]

97. Chow, B.Y.; Han, X.; Dobry, A.S.; Qian, X.; Chuong, A.S.; Li, M.; Henninger, M.A.; Belfort, G.M.; Lin, Y.; Monahan, P.E. High-performance genetically targetable optical neural silencing by light-driven proton pumps. *Nature* **2010**, *463*, 98–102. [CrossRef] [PubMed]
98. Tsunoda, S.P.; Ewers, D.; Gazzarrini, S.; Moroni, A.; Gradmann, D.; Hegemann, P. H^+-pumping rhodopsin from the marine alga Acetabularia. *Biophys. J.* **2006**, *91*, 1471–1479. [CrossRef]
99. Lou, S.; Adam, Y.; Weinstein, E.N.; Williams, E.; Williams, K.; Parot, V.; Kavokine, N.; Liberles, S.; Madisen, S.; Zeng, H. Genetically Targeted All-Optical Electrophysiology with a Transgenic Cre-Dependent Optopatch Mouse. *J. Neurosci.* **2016**, *36*, 11059–11073. [CrossRef]
100. Grimm, J.B.; English, B.P.; Chen, J.; Slaughter, J.P.; Zhang, Z.; Revyakin, A.; Patel, R.; Macklin, J.J.; Normanno, D.A.; Singer, R.H. A general method to improve fluorophores for live-cell and single-molecule microscopy. *Nat. Methods* **2015**. [CrossRef]
101. Lim, S.T.; Antonucci, D.E.; Scannevin, R.H.; Trimmer, J.S. A Novel Targeting Signal for Proximal Clustering of the Kv2.1 K^+ Channel in Hippocampal Neurons. *Neuron* **2000**, *25*, 385–397. [CrossRef]
102. Baker, C.A.; Elyada, Y.M.; Parra, A.; Bolton, M.M.L. Cellular resolution circuit mapping with temporal-focused excitation of soma-targeted channelrhodopsin. *Elife* **2016**, *5*, 1–15. [CrossRef] [PubMed]
103. Shemesh, O.; Tanese, D.; Zampini, V.; Changyang, L.; Kiryln, P.; Ronzitti, E.; Papagiakoumou, E.; Boyden, E.S.; Emiliani, V. Temporally precise single-cell resolution optogenetics. *Nat. Neurosci.* **2017**, *20*, 1796. [CrossRef] [PubMed]
104. Chien, M.-P.; Brinks, D.; Adam, Y.; Bloxham, W.; Kheifets, S.; Cohen, A.E. Two-photon photoactivated voltage imaging in tissue with an Archaerhodopsin-derived reporter. *bioRxiv* **2017**. [CrossRef]
105. Cao, G.; Platisa, J.; Pieribone, V.A.; Raccuglia, D.; Kunst, M.; Nitabach, M.N. Genetically targeted optical electrophysiology in intact neural circuits. *Cell* **2013**, *154*, 904–913. [CrossRef]
106. Yang, H.H.H.; St-Pierre, F.; Sun, X.; Ding, X.; Lin, M.Z.Z.; Clandinin, T.R.R. Subcellular Imaging of Voltage and Calcium Signals Reveals Neural Processing In Vivo. *Cell* **2016**, *166*, 245–257. [CrossRef] [PubMed]
107. Chamberland, S.; Yang, H.H.; Pan, M.M.; Evans, S.W.; Guan, S.; Chavarha, M.; Yang, Y.; Salesse, C.; Wu, H.; Wu, J.C. Fast two-photon imaging of subcellular voltage dynamics in neuronal tissue with genetically encoded indicators. *Elife* **2017**, *6*, 1–35. [CrossRef]
108. Roome, C.J.; Kuhn, B. Simultaneous dendritic voltage and calcium imaging and somatic recording from Purkinje neurons in awake mice. *Nat. Commun.* **2018**, *9*, 1–14. [CrossRef]
109. Kulkarni, R.U.; Vandenberghe, M.; Thunemann, M.; James, F.; Andreassen, O.A.; Djurovic, S.; Devor, A.; Miller, E.W. In Vivo Two-Photon Voltage Imaging with Sulfonated Rhodamine Dyes. *ACS Cent. Sci.* **2018**, *4*, 1371–1378. [CrossRef]
110. St-Pierre, F.; Marshall, J.D.; Yang, Y.; Gong, Y.; Schnitzer, M.J.; Lin, M.Z. High-fidelity optical reporting of neuronal electrical activity with an ultrafast fluorescent voltage sensor. *Nat. Neurosci.* **2014**, *17*, 884–889. [CrossRef]
111. Südhof, T.C. Neurotransmitter release: The last millisecond in the life of a synaptic vesicle. *Neuron* **2013**. [CrossRef] [PubMed]
112. Scimemi, A.; Beato, M. Determining the neurotransmitter concentration profile at active synapses. *Mol. Neurobiol.* **2009**. [CrossRef]
113. Lüscher, C.; Malenka, R.C. NMDA receptor-dependent long-term potentiation and long-term depression (LTP/LTD). *Cold Spring Harb. Perspect. Biol.* **2012**. [CrossRef] [PubMed]
114. Von Gersdorff, H.; Sakaba, T.; Berglund, K.; Tachibana, M. Submillisecond kinetics of glutamate release from a sensory synapse. *Neuron* **1998**. [CrossRef]
115. Meldrum, B.S. Glutamate as a Neurotransmitter in the Brain: Review of Physiology and Pathology. *J. Nutr.* **2000**, *130*, 1007S–1015S. [CrossRef] [PubMed]
116. McCormick, D.A. GABA as an inhibitory neurotransmitter in human cerebral cortex. *J. Neurophysiol.* **1989**, *62*, 1018–1027. [CrossRef] [PubMed]
117. Rudnick, G. *Neurotransmitter Transporters: Structure, Function, and Regulation*; Reith, M.E.A., Ed.; Humana Press: New York, NY, USA, 2002; pp. 25–52.
118. Snyder, S.H. Brain peptides as neurotransmitters. *Science* **1980**, *209*, 976–983. [CrossRef]
119. Kreitzer, A.C. Neurotransmission: Emerging Roles of Endocannabinoids. *Curr. Biol.* **2005**, *15*, R549–R551. [CrossRef]

120. Dawson, T.M.; Snyder, S.H. Gases as biological messengers: Nitric oxide and carbon monoxide in the brain. *J. Neurosci.* **1994**, *14*, 5147–5159. [CrossRef]
121. Ames, G.F.L. Bacterial Periplasmic Transport Systems: Structure, Mechanism, and Evolution. *Annu. Rev. Biochem.* **1986**. [CrossRef]
122. De Lorimier, R.M.; Smith, J.J.; Dwyer, M.A.; Looger, L.L.; Sali, K.M.; Paavola, C.D.; Rizk, S.S.; Sadigov, S.; Conrad, D.W.; Loew, L. Construction of a fluorescent biosensor family. *Protein Sci.* **2002**, *11*, 2655–2675. [CrossRef] [PubMed]
123. Dwyer, M.A.; Hellinga, H.W. Periplasmic binding proteins: A versatile superfamily for protein engineering. *Curr. Opin. Struct. Biol.* **2004**, *14*, 495–504. [CrossRef] [PubMed]
124. Deuschle, K.; Okumoto, S.; Fehr, M.; Looger, L.L.; Kozhukh, L.; Frommer, W.B. Construction and optimization of a family of genetically encoded metabolite sensors by semirational protein engineering. *Protein Sci.* **2005**, *14*, 2304–2314. [CrossRef] [PubMed]
125. Okumoto, S.; Looger, L.L.; Micheva, K.D.; Reimer, R.J.; Smith, S.J.; Frommer, W.B. Detection of glutamate release from neurons by genetically encoded surface-displayed FRET nanosensors. *Proc. Natl. Acad. Sci. USA* **2005**, *102*, 8740–8745. [CrossRef] [PubMed]
126. Hires, S.A.; Zhu, Y.; Tsien, R.Y. Optical measurement of synaptic glutamate spillover and reuptake by linker optimized glutamate-sensitive fluorescent reporters. *Proc. Natl. Acad. Sci. USA* **2008**, *105*, 4411–4416. [CrossRef] [PubMed]
127. Marvin, J.S.; Borghuis, B.G.; Tian, L.; Cichon, J.; Harnett, M.T.; Akerboom, J.; Gordus, A.; Renninger, S.L.; Chen, T.W.; Bargmann, C.I. An optimized fluorescent probe for visualizing glutamate neurotransmission. *Nat. Methods* **2013**, *10*, 162–170. [CrossRef] [PubMed]
128. Taschenberger, H.; Woehler, A.; Neher, E. Superpriming of synaptic vesicles as a common basis for intersynapse variability and modulation of synaptic strength. *Proc. Natl. Acad. Sci. USA* **2016**. [CrossRef]
129. Helassa, N.; Dürst, C.D.; Coates, C.; Kerruth, S.; Arif, U.; Schulze, C.; Wiegert, J.S.; Geeves, M.; Oertner, T.G.; Török, K. Ultrafast glutamate sensors resolve high-frequency release at Schaffer collateral synapses. *Proc. Natl. Acad. Sci. USA* **2018**, *115*, 5594–5599. [CrossRef]
130. Marvin, J.S.; Scholl, B.; Wilson, D.E.; Podgorski, K.; Kazemipour, A.; Müller, J.A.; Schoch, S.; Quiroz, F.J.U.; Rebola, N.; Bao, H. Stability, affinity, and chromatic variants of the glutamate sensor iGluSnFR. *Nat. Methods* **2018**, *15*, 936–939. [CrossRef]
131. Wu, J.; Abdelfattah, A.S.; Zhou, H.; Ruangkittisakul, A.; Qian, Y.; Ballanyi, K.; Campbell, R.E. Genetically Encoded Glutamate Indicators with Altered Color and Topology. *ACS Chem. Biol.* **2018**, *13*, 1832–1837. [CrossRef]
132. Looger, L.L.; Marvin, J.S.; Shimoda, Y.; Magloire, V.; Leite, M.; Kawashima, T.; Jensen, T.P.; Knott, E.L.; Novak, O.; Podgorski, K. A genetically encoded fluorescent sensor for in vivo imaging of GABA. *bioRxiv* **2018**. [CrossRef]
133. Zhang, W.H.; Herde, M.K.; Mitchell, J.A.; Whitfield, J.H.; Wulff, A.B.; Vongsouthi, V.; Sanchez-Romero, I.; Gulakova, P.E.; Minge, D.; Breithausen, B. Monitoring hippocampal glycine with the computationally designed optical sensor GlyFS. *Nat. Chem. Biol.* **2018**, *14*, 861–869. [CrossRef]
134. Patriarchi, T.; Cho, J.R.; Merten, K.; Howe, M.W.; Marley, A.; Xiong, W.H.; Folk, R.W.; Broussard, G.J.; Liang, R.; Jang, M.J. Ultrafast neuronal imaging of dopamine dynamics with designed genetically encoded sensors. *Science* **2018**, *360*, 1–15. [CrossRef] [PubMed]
135. Sun, F.; Zeng, J.; Jing, M.; Zhou, J.; Feng, J.; Owen, S.F.; Luo, Y.; Li, F.; Wang, H.; Yamaguchi, T. A Genetically Encoded Fluorescent Sensor Enables Rapid and Specific Detection of Dopamine in Flies, Fish, and Mice. *Cell* **2018**, *174*, 481–496. [CrossRef] [PubMed]
136. Jing, M.; Zhang, P.; Wang, G.; Feng, J.; Mesik, L.; Zeng, J.; Jiang, H.; Wang, S.; Looby, J.C.; Guagliardo, N.A. A genetically encoded fluorescent acetylcholine indicator for in vitro and in vivo studies. *Nat. Biotechnol.* **2018**, *36*, 726–737. [CrossRef] [PubMed]
137. Feng, J.; Zhang, C.; Lischinsky, J.; Jing, M.; Zhou, J.; Wang, H.; Zhang, Y.; Dong, A.; Wu, Z.; Wu, H.; et al. A genetically encoded fluorescent sensor for rapid and specific in vivo detection of norepinephrine. *bioRxiv* **2018**. [CrossRef]
138. Lee, D.; Creed, M.; Jung, K.; Stefanelli, T.; Wendler, D.J.; Oh, W.C.; Mignocchi, N.L.; Lüscher, C.; Kwon, H.B. Temporally precise labeling and control of neuromodulatory circuits in the mammalian brain. *Nat. Methods* **2017**, *14*, 495–503. [CrossRef] [PubMed]

139. Guthrie, G.D.; Nicholson-Guthrie, C.S. gamma-Aminobutyric acid uptake by a bacterial system with neurotransmitter binding characteristics. *Proc. Natl. Acad. Sci. USA* **1989**. [CrossRef]
140. Planamente, S.; Vigouroux, A.; Mondy, S.; Nicaise, M.; Faure, D.; Moréra, S. A conserved mechanism of GABA binding and antagonism is revealed by structure-function analysis of the periplasmic binding protein Atu2422 in Agrobacterium tumefaciens. *J. Biol. Chem.* **2010**. [CrossRef]
141. Whitfield, J.H.; Zhang, W.H.; Herde, M.K.; Clifton, B.E.; Radziejewski, J.; Janovjak, H.; Henneberger, C.; Jackson, C.J. Construction of a robust and sensitive arginine biosensor through ancestral protein reconstruction. *Protein Sci.* **2015**. [CrossRef]
142. Tannous, B.A.; Grimm, J.; Perry, K.F.; Chen, J.W.; Weissleder, R.; Breakefield, X.O. Metabolic biotinylation of cell surface receptors for in vivo imaging. *Nat. Methods* **2006**. [CrossRef] [PubMed]
143. Park, J.H.; Scheerer, P.; Hofmann, K.P.; Choe, H.W.; Ernst, O.P. Crystal structure of the ligand-free G-protein-coupled receptor opsin. *Nature* **2008**. [CrossRef] [PubMed]
144. Scheerer, P.; Park, J.H.; Hildebrand, P.W.; Kim, Y.J.; Krauß, N.; Choe, H.W.; Hofmann, K.P.; Ernst, O.P. Crystal structure of opsin in its G-protein-interacting conformation. *Nature* **2008**. [CrossRef] [PubMed]
145. Vilardaga, J.P.; Bünemann, M.; Krasell, C.; Castro, M.; Lohse, M.J. Measurement of the millisecond activation switch of G protein-coupled receptors in living cells. *Nat. Biotechnol.* **2003**, *21*, 807–812. [CrossRef]
146. Ziegler, N.; Bätz, J.; Zabel, U.; Lohse, M.J.; Hoffmann, C. FRET-based sensors for the human M1-, M3-, and M 5-acetylcholine receptors. *Bioorganic Med. Chem.* **2011**, *19*, 1048–1054. [CrossRef] [PubMed]
147. Masharina, A.; Reymond, L.; Maurel, D.; Umezawa, K.; Johnsson, K. A Fluorescent Sensor for GABA and Synthetic GABA. *J. Am. Chem. Soc.* **2012**, *134*, 19026–19034. [CrossRef] [PubMed]
148. Maier-Peuschel, M.; Frölich, N.; Dees, C.; Hommers, L.G.; Hoffmann, C.; Nikolaev, V.O.; Lohse, M.J. A fluorescence resonance energy transfer-based M2 muscarinic receptor sensor reveals rapid kinetics of allosteric modulation. *J. Biol. Chem.* **2010**, *285*, 8793–8800. [CrossRef] [PubMed]
149. Luttrell, L.M.; Lefkowitz, R.J. The role of ß-arrestins in the termination and transduction of G-protein-coupled receptor signals. *J. Cell Sci.* **2002**. [CrossRef]
150. Neves, S.R.; Ram, P.T.; Iyengar, R. G protein pathways. *Science* **2002**. [CrossRef] [PubMed]
151. Barnea, G.; Strapps, W.; Herrada, G.; Berman, Y.; Ong, J.; Kloss, B.; Axel, R.; Lee, K.J. The genetic design of signaling cascades to record receptor activation. *Proc. Natl. Acad. Sci. USA* **2008**, *105*, 64–69. [CrossRef]
152. Nguyen, Q.T.; Schroeder, L.F.; Mank, M.; Muller, A.; Taylor, P.; Griesbeck, O.; Kleinfeld, D. An in vivo biosensor for neurotransmitter release and in situ receptor activity. *Nat. Neurosci.* **2010**, *13*, 127–132. [CrossRef] [PubMed]
153. Muller, A.; Joseph, V.; Slesinger, P.A.; Kleinfeld, D. Cell-based reporters reveal in vivo dynamics of dopamine and norepinephrine release in murine cortex. *Nat. Methods* **2014**, *11*, 1245–1252. [CrossRef] [PubMed]
154. Guntas, G.; Hallett, R.A.; Zimmerman, S.P.; Williams, T.; Yumerefendi, H.; Bear, J.E.; Kuhlman, B. Engineering an improved light-induced dimer (iLID) for controlling the localization and activity of signaling proteins. *Proc. Natl. Acad. Sci. USA* **2015**. [CrossRef] [PubMed]
155. Overbeek, R.; Bartels, D.; Vonstein, V.; Meyer, F. Annotation of Bacterial and Archaeal Genomes: Improving Accuracy and Consistency. *Chem. Rev.* **2007**, *107*, 3431–3447. [CrossRef] [PubMed]
156. Markowitz, V.M.; Chen, I.-M.A.; Chu, K.; Pati, A.; Ivanova, N.N.; Kyrpides, N.C. Ten Years of Maintaining and Expanding a Microbial Genome and Metagenome Analysis System. *Trends Microbiol.* **2015**, *23*, 730–741. [CrossRef] [PubMed]
157. Packer, M.S.; Liu, D.R. Methods for the directed evolution of proteins. *Nat. Rev. Genet.* **2015**, *16*, 379–394. [CrossRef] [PubMed]
158. Subach, F.V.; Piatkevich, K.D.; Verkhusha, V.V. Directed molecular evolution to design advanced red fluorescent proteins. *Nat. Methods* **2011**. [CrossRef]
159. Chia, T.H.; Levene, M.J. Microprisms for in vivo multilayer cortical imaging. *J. Neurophysiol.* **2009**, *102*, 1310–1314. [CrossRef]
160. Andermann, M.L.; Gilfoy, N.B.; Goldey, G.J.; Sachdev, R.N.S.; Wölfel, M.; McCormick, D.A.; Reid, R.C.; Levene, M.J. Chronic Cellular Imaging of Entire Cortical Columns in Awake Mice Using Microprisms. *Neuron* **2013**, *80*, 900–913. [CrossRef]
161. Ghosh, K.K.; Burns, L.D.; Cocker, E.D.; Nimmerjahn, A.; Ziv, Y.; El Gamal, A.; Schnitzer, M.J. Miniaturized integration of a fluorescence microscope. *Nat. Methods* **2011**, *8*, 871–878. [CrossRef]

162. Yu, K.; Ahrens, S.; Zhang, X.; Schiff, H.; Ramakrishnan, C.; Fenno, L.; Deisseroth, K.; Zhao, F.; Luo, M.H.; Gong, L. The central amygdala controls learning in the lateral amygdala. *Nat. Neurosci.* **2017**, *20*, 1680. [CrossRef] [PubMed]
163. Tillberg, P.W.; Chen, F.; Piatkevich, K.D.; Zhao, Y.; Yu, C.-C.; English, B.P.; Gao, L.; Martorell, A.; Suk, H.J.; Yoshida, F. Protein-retention expansion microscopy of cells and tissues labeled using standard fluorescent proteins and antibodies. *Nat. Biotechnol.* **2016**, *34*, 987–992. [CrossRef] [PubMed]
164. Wassie, A.T.; Zhao, Y.; Boyden, E.S. Expansion microscopy: Principles and uses in biological research. *Nat. Methods* **2019**, *16*, 33–41. [CrossRef] [PubMed]
165. Lee, K.-S.; Vandemark, K.; Mezey, D.; Shultz, N.; Fitzpatrick, D. Functional Synaptic Architecture of Callosal Inputs in Mouse Primary Visual Cortex. *Neuron* **2018**, 1–8. [CrossRef]

Article

Engineering Optogenetic Control of Endogenous p53 Protein Levels

Pierre Wehler [1,2] and Barbara Di Ventura [1,2,*]

1 Faculty of Biology, University of Freiburg, 79104 Freiburg, Germany; pierre.wehler@biologie.uni-freiburg.de
2 Signalling Research Centres BIOSS and CIBSS, University of Freiburg, 79104 Freiburg, Germany
* Correspondence: barbara.diventura@biologie.uni-freiburg.de

Received: 2 May 2019; Accepted: 20 May 2019; Published: 21 May 2019

Featured Application: This study describes the development of an optogenetic tool to control endogenous p53 levels with blue light.

Abstract: The transcription factor p53 is a stress sensor that turns specific sets of genes on to allow the cell to respond to the stress depending on its severity and type. p53 is classified as tumor suppressor because its function is to maintain genome integrity promoting cell cycle arrest, apoptosis, or senescence to avoid proliferation of cells with damaged DNA. While in many human cancers the p53 gene is itself mutated, there are some in which the dysfunction of the p53 pathway is caused by the overexpression of negative regulators of p53, such as Mdm2, that keep it at low levels at all times. Here we develop an optogenetic approach to control endogenous p53 levels with blue light. Specifically, we control the nuclear localization of the Mmd2-binding PMI peptide using the light-inducible export system LEXY. In the dark, the PMI-LEXY fusion is nuclear and binds to Mdm2, consenting to p53 to accumulate and transcribe the target gene p21. Blue light exposure leads to the export of the PMI-LEXY fusion into the cytosol, thereby Mdm2 is able to degrade p53 as in the absence of the peptide. This approach may be useful to study the effect of localized p53 activation within a tissue or organ.

Keywords: Optogenetics; p53; *As*LOV2; LINuS; LEXY; MIP; PMI

1. Introduction

Despite its unassuming name, p53 is an essential protein for the cell and is evolutionarily conserved from worm to human [1,2]. Acting as a sensor of a variety of stress signals, p53 dictates life or death to the cell, depending on whether it is able or not to cope with the stress and return to homeostasis [3]. The fundamental role of p53 in ensuring genomic integrity is better reflected in its nickname, "guardian of the genome". This and other epithets, such as "good cop/bad cop" or "heavily dictated dictator", emphasize not only its centrality in keeping proliferating cells under control, but also suggest that p53 can have contradicting effects on the organism (like cancer prevention and aging) and is therefore itself subjected to a tight regulation achieved by a complex cellular network [4,5]. p53 is mainly a sequence-specific DNA-binding protein which, under normal growth conditions, is kept latent in various ways [5,6]. The most prominent mechanism to keep p53 latent when not needed is to maintain it at very low levels. This is achieved by a series of E3 ubiquitin ligases that target p53 for proteasomal degradation by ubiquitylating it on lysine residues [7,8]. The oncoprotein murine double minute 2 (Mdm2) appears to be the major negative regulator of p53 [9,10]. Beyond targeting it for degradation, Mdm2 inhibits p53 also by promoting its nuclear export [9] and by binding to its N-terminal transactivation domain (TAD), which is therefore not available for binding to the DNA [11–13]. Interestingly, the Mdm2 homologous protein, murine double minute 4 (MdmX), inhibits p53 via the same TAD-sequestration strategy, but lacks the ability to promote p53 degradation, despite having a RING domain like Mdm2 [14].

When necessity calls, i.e., upon stress signals such as DNA damage, hypoxia, or deregulated expression of oncogenes, p53 is rapidly activated via acquisition of post-translational modifications, some of which enable its stabilization and the consequent increase in its protein levels [5]. This post-translationally modified, stabilized p53 can then start its gene expression program. p53 target genes mediate a plethora of cellular functions, including cell cycle arrest, senescence, apoptosis, differentiation, and DNA repair, to name but a few [2].

Inactivation of the p53 pathway—be it through mutations of the p53 gene itself, mutations of genes encoding key p53 regulators, and/or binding partners, or other means—is a common, if not universal feature of human cancers. The cancers in which p53 is not mutated, but rather aberrantly kept at low levels at all times by means of overexpressed negative regulators [15,16], have better chances of being defeated because, if p53 levels could be brought back to normal, its target genes could be activated. Indeed, reactivation of the endogenous p53 signaling pathway has been shown to lead to tumor regression [17,18]. Therefore, disrupting an aberrant p53-Mdm2/MdmX interaction on a wild type p53 background could be a promising therapy.

Another feature that would make a new therapy promising would be the ability to target it only at the cancer cells. Indeed, the major drawback of conventional chemotherapy is the lack of specific distribution of the drugs to the interested tissues and organs. The drugs distribute non-specifically inside the body, leading to substantial damage of healthy cells alongside the diseased ones. To overcome this problem, new concepts have been developed, such as delivery of chemotherapeutic drugs from vehicles (nanoparticles, liposomes, polymeric micells, minicells, etc.) that accumulate inside cancer cells for instance via ligand-receptor interactions [19] or that release their content locally upon illumination with near-infrared (NIR) light [20,21].

Alongside its use in combination with chemical compounds, light has been exploited to regulate genetically encoded photosensors within cells, giving rise to the field of optogenetics [22,23]. One of the workhorses of optogenetics is the second light-oxygen-voltage (LOV) domain of *Avena sativa* phototrophin 1 (*As*LOV2 domain). In the dark, the C-terminal Jα helix of the *As*LOV2 domain is folded and bound to the core domain [24–28] once the chromophore flavin mononucleotide (FMN) absorbs photons of ~450 nm light, a structural rearrangement occurs within the protein triggering the unfolding and undocking of the Jα helix [24–28]. The recovery to the inactive dark state occurs spontaneously within less than a minute [29]. This conformational change has been harnessed to control, with light, the exposure of diverse peptides that have been appended to the Jα helix [30–32]. In particular, nuclear localization and export signals (NLSs and NESs) were photocaged within *As*LOV2 giving rise to the optogenetic tools named LINuS [33] and LANS [34] for controlling nuclear protein import, and LEXY [35] and LINX [36] for controlling nuclear protein export.

Here we explore the possibility to use optogenetics to elevate p53 levels by disrupting its interaction with Mdm2 and MdmX. We compare two previously published peptides, MIP [37] and PMI [38], and find that, under our experimental conditions, only PMI results in higher p53 protein levels in HCT116 cells. Interestingly, we discover that the peptide needs to localize to the nucleus to exert its function. Therefore, we fuse it to LEXY and show that the PMI-mCherry-LEXY construct allows controlling with light endogenous p53 levels in HCT116 cells. Specifically, cells kept in the dark have higher p53 levels than cells exposed to blue light which triggers export of the PMI-mCherry-LEXY construct into the cytosol, thereby preventing PMI to exert its function. Stabilized p53 leads to the accumulation of the p53 target gene p21 [39,40] in HCT116 cells kept in the dark. Even if not applicable in a therapeutic scenario in the present form—as cells with elevated p53 levels are, in this case, those left in the dark—this approach should be useful to study the effect of having a heterogeneous p53 status within a tissue or organ.

2. Materials and Methods

2.1. Plasmid Generation

The plasmids expressing either p53 or thioredoxin bearing the MIP/PMI/3A peptide fused to mCherry were constructed from pcDNA3.1(+) (ThermoFischer Scientific) by first inserting the mCherry coding sequence between the BamHI and XhoI sites of the multi-cloning site (MCS) of pcDNA3.1(+) and then inserting the p53 or thioredoxin coding sequence into this modified vector between the HindIII and BamHI sites. The NLS/NES upstream of mCherry was cloned by adding the corresponding sequence directly into the forward primer used to amplify the *mcherry* gene. Peptides were inserted as oligos into thioredoxin between Gly33 and Pro34 using golden gate cloning. Addition of PMI-linker(GGS)-NLS sequences to mCherry, LINuS, or LEXY constructs was done using overhang PCR. All constructs were verified by DNA sequencing. The LINuS and LEXY constructs used here have been previously described [33,35].

2.2. Illumination of Cells with Blue Light

Cells were illuminated with 20 μmol s^{-1} m^{-2} of 480 nm light within the cell culture incubator using a custom-made LED device connected to a power box (Manson HCS-3102) controlled by a Raspberry Pi.

2.3. Cell Culture and Transient Transfection

The human colon carcinoma cell line HCT116 was kindly provided by Thomas Hofmann, University of Mainz. HCT116 cells were maintained in phenol red-free Dulbecco's Modified Eagle Medium supplemented with 10% FCS (Sigma), 2 mM L-glutamine (Life Technologies) and 1% PenStrep (Life Technologies). Cells were cultured at 37 °C and 5% CO_2 in a humidified tissue culture incubator. Cells were transfected with 2500 ng total DNA (250 ng construct DNA and 2250 ng empty pcDNA3.1 (+)) using Lipofectamin 2000 according to the manufactures protocol. Mock transfected samples were transfected with empty pcDNA3.1 (+).

2.4. Western Blot

Cells were seeded into 6-well plates and transfected as described above. For non-optogenetics experiments, cells were kept in the dark at all times and lysed 32 h post-transfection. For optogenetics experiments, cells were incubated right after transfection under blue light (20 μmol s^{-1} m^{-2}; 480 nm) overnight. The next day, one plate was incubated in the dark (dark control), while the other plate(s) remained under illumination. 24 h later, cells were collected in ice-cold lysis buffer (20 mM Tris-HCl pH 7.4, 1% Triton X.100, 10% glycerol, 150 mM NaCl, 1% phenylmethylsulfonyl fluoride, 1% benzonase (Novagen, 70664), and 1 Complete Mini Protease Inhibitor tablet (Roche, 11 836 153 001)). In the experiment shown in Figure 1b,c, the double-strand DNA breaks-inducing anthracycline antibiotic daunorubicin was added at a final concentration of 0.5 μM 8 h after transfection. Protein concentration was measured by Bradford assay and adjusted to 1 μg\mL. 15 μg were loaded on a 12% Bis-Tris gel and proteins were separated by electrophoresis. Proteins were then transferred onto a polyvinylidene difluoride membrane, which was blocked using 5% BSA in PBS-Tween 20 (PBS-T). Primary antibodies were diluted in 5% BSA in PBS-T and applied for 1 h to detect p53 (Santa Cruz, sc-126, diluted 1:1000) or p21 (BD Pharmigen, 556430, diluted 1:666) and beta-actin (Abcam, ab8226, diluted 1:1000), followed by incubation with a secondary goat anti-mouse IgG(H+L)-PRP0 (Dianova, 115-035-003) for 45 min. Chemiluminescence was detected using the SuperSignal West Pico Chemiluminescent Substrate (Thermo Scientific) and the ChemoCam Imager (Intas).

2.5. Confocal Imaging

Cells were imaged at 37 °C and 5% CO_2 in ibidi μ-slide 8 wells (ibidi, 8226) on a Leica Sp5 confocal microscope equipped with an incubation chamber and a 20x air objective (0.7 NA). Activation was performed imaging cells with the 458 nm laser at 80% intensity for 5 seconds every 30 s for a total of 40 min. mCherry was excited using the 561 nm laser. Cells were focused with the 561 nm laser to prevent premature activation of LINuS/LEXY.

2.6. Data Processing

2.6.1. Quantification of Protein Levels

Western blots were quantified by measuring band intensities using the ImageJ Gels package. The intensity of the band corresponding to the loading control was used to normalize the intensity of the band corresponding to the protein under investigation (p53 or p21).

2.6.2. Image Analysis

Microscopy images were processed using ImageJ. First, the background was subtracted, then a circular ROI was drawn manually in the nuclei of all cells in a given field of view. The mean intensity of the nuclei was quantified at each time point. Values are normalized to that at time point zero (before illumination started).

2.6.3. Statistical Analysis

Independent replicates refer to independent cell samples of experiments carried out on different days. If not indicated otherwise, bars and graph represent the mean values, the error bars the standard deviation of the mean. Statistical analysis was done by a two-tailed unpaired Student's *t*-test; P values (p) < 0.05 (*), < 0.01 (**), and < 0.001 (***) were considered statistically significant. P values ≥ 0.05 were considered statistically not significant (NS).

3. Results

3.1. The Mdm2-inhibitory Peptide (MIP) Does Not Elevate p53 Levels in HCT116 Cells

We started by testing the Mdm2-inhibitory peptide MIP (PRFWEYWLRLME), which was selected *in vitro* from large libraries of random peptides using mRNA display [37]. MIP binds Mdm2 and also MdmX with higher affinity than other known peptides, such as DI [41]. We cloned the MIP peptide into a freely accessible loop of *Escherichia coli* thioredoxin, a small and stable protein commonly used as scaffold for presenting and stabilizing peptides. mCherry was added to the construct to consent easy determination of transfection efficiency.

As a control, we cloned the 3A peptide (LTAEHYAAQATS) in which three key hydrophobic residues of the DI peptide are mutated to alanine, thereby abrogating binding to Mdm2 or MdmX.

To test the effect of the peptide we selected HCT116 cells, a human colon cancer cell line with wild type p53, which has been shown to accumulate upon inhibition of Mdm2 [42]. Expression of Trx-MIP-mCherry did not lead to changes in p53 protein levels, which remained similar to those in mock and Trx-3A-mCherry transfected samples (Supplementary Figure S1). This is in contrast with previously published results [37]. The raw data used to quantify protein levels are found in the Supplementary Materials (Figures S2–S5).

3.2. The p53-Mdm2/MdmX Inhibitor (PMI) Peptide Elevates p53 Levels in HCT116 Cells

We then turned to another peptide, the so-called p53-Mdm2/MdmX inhibitor (PMI) peptide [38] (TSFAEYWNLLSP). PMI was identified in a phage display, using as baits the p53-binding domains of Mdm2 and MdmX. We followed the same approach as for the MIP peptide, thus cloning PMI into the thioredoxin scaffold fused to mCherry, and then testing its function in HCT116 cells.

36 h after transfection, cells expressing Trx-PMI-mCherry exhibited about 60% increase in p53 levels compared to the mock transfected cells, implying that degradation of p53 was inhibited (Figure 1a). As seen in previous experiments, expression of Trx-3A-mCherry and Trx-MIP-mCherry did not affect p53 levels.

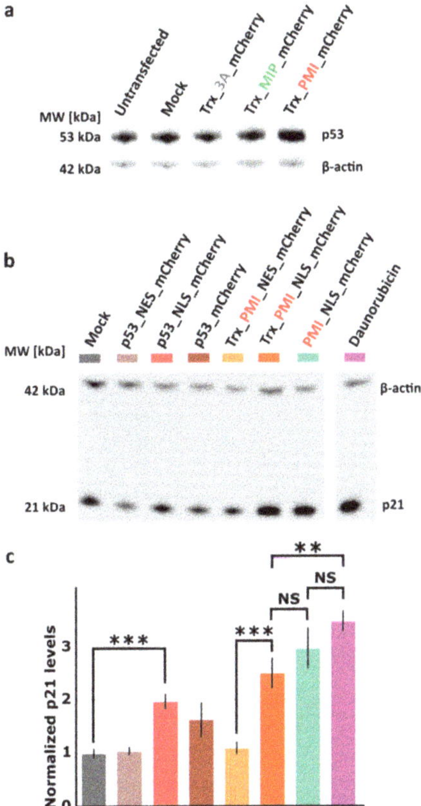

Figure 1. The p53-MDM2/MDMX inhibitor (PMI) peptide stabilizes p53 in HCT116 cells in a localization-dependent manner. (a), (b) Western blot showing the levels of p53 (a) and p21 (b) in HCT116 cells transfected with the indicated constructs. Mock indicates cells transfected with empty pcDNA3.1(+). β-actin was used as loading control. (**a**) This experiment was conducted once. (**b**) Daunorubicin was added at 0.5 µM 8 h after transfection. Cells were collected for Western blot analysis after 24 h. A lane with a sample not relevant to this study was cut from the image and is shown as a white space. (**c**) Bar plot showing the quantification of two independent Western blots with samples as in (b). Data indicate mean ± standard deviation. Color code as in (b). NS $p \geq 0.05$, ** $p < 0.01$, *** $p < 0.001$, *t*-test.

Next, we asked whether PMI had to localize to either the nucleus or the cytosol to exert its function. For this purpose, we added a nuclear localization signal (NLS) or a nuclear export signal (NES) to the Trx-PMI-mCherry construct. This time, we decided to look directly at the effect of the peptide on one of p53 target genes, p21. Indeed, it is important to make sure that stabilized p53 is fully functional. As additional controls beyond the mock transfected cells, we included cytoplasmic p53 (p53-NES-mCherry), nuclear p53 (p53-NLS-mCherry and p53-mCherry, which is largely nuclear even in the absence of an additional NLS) and daunorubicin, a double-strand DNA breaks-inducing drug. Interestingly, we found that the peptide had to be nuclear to elevate p21 levels (Figure 1b,c).

We then tested if we could dispense of the thioredoxin scaffold by expressing PMI simply as a fusion to mCherry. This would further reduce the size of the final construct. We found that the PMI-mCherry fusion worked as well as the construct based on thioredoxin, giving rise to p53 levels similar to those obtained when using daunorubicin (Figure 1b,c).

3.3. Developing Optogenetic Control of PMI Localization

As the function of PMI is dependent on its nuclear localization, we reasoned that, if we could control this latter with LINuS, we could effectively control endogenous p53 levels with blue light. To this aim, we cloned three constructs differing only in the LINuS variant used, as different NLSs photocaged in the Jα helix of the *As*LOV2 domain have different properties such as background in the dark or extent of nuclear accumulation after light induction [33] (Figure 2a). Albeit two of the three constructs showed statistically significant nuclear accumulation after light induction (Figure 2b,e), we were not satisfied with their performance, as the fold change in nuclear localization before and after light was less than two.

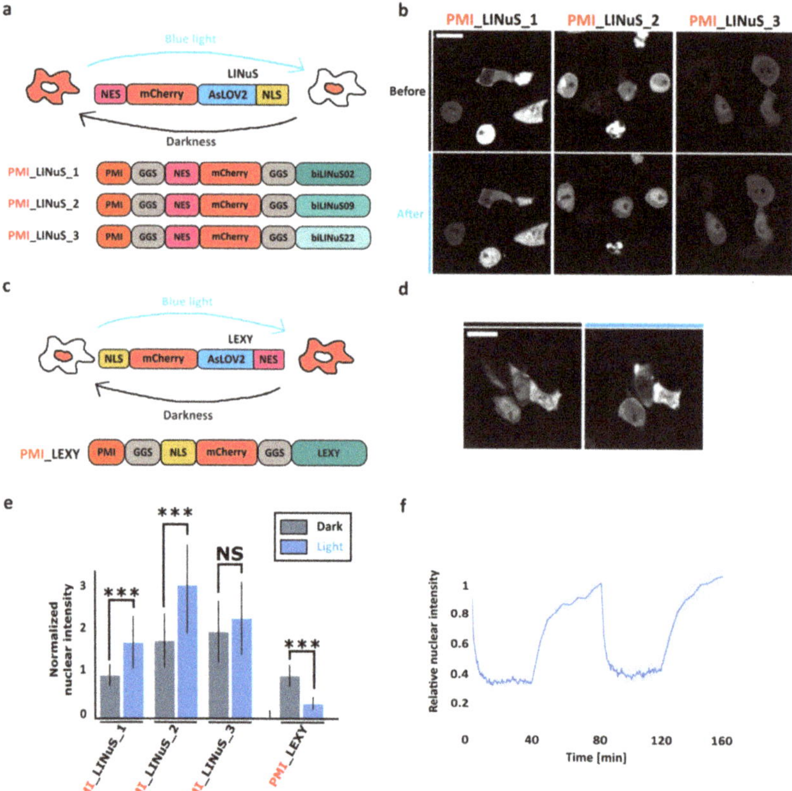

Figure 2. Engineering optogenetic control of PMI localization. (**a**,**c**) Schematic representation of the PMI_LINuS and PMI_LEXY constructs. The NES is the PKIt NES. GGS is a flexible linker. (**b**,**d**) Representative images of HCT116 cells transiently transfected with the indicated constructs before and after 40 min of blue light illumination. (**b**,**d**) Scale bar, 20 µm. (**e**) Bar plot showing the nuclear intensity of the indicated constructs in the dark or after illumination with blue light normalized to

the values for the PMI_LINuS_1 construct in the dark (for the LINuS constructs) and to PMI_LEXY construct in the dark (for the LEXY construct). NS $p \geq 0.05$, *** $p < 0.001$, *t*-test. (**f**) Plot showing the nuclear intensity of the mCherry fluorescence measured in HCT116 cells transiently transfected with the PMI_LEXY construct shown in (**c**) in cells illuminated with blue light twice with a 40 min-long activation phase. Graph represents mean plus 95% confidence interval. N = 23 cells.

Therefore, we turned to LEXY, which is similar to LINuS in that it is also based on *As*LOV2, however, in this case, the PMI peptide would be nuclear unless light is applied to the cells. LEXY has proven to be easily applicable to various proteins of interest without the need of optimization [35,43,44]. PMI fused to mCherry and LEXY (PMI-NLS-mCherry-LEXY, for simplicity referred to as PMI_LEXY from now on) could be robustly accumulated into the cytoplasm after blue light illumination of HCT116 cells (Figure 2c–e), showing a fold change close to three. We further confirmed the reversibility of this accumulation (Figure 2f).

3.4. PMI_LEXY Can Be Used to Control with Light Endogenous p53 and p21 Levels

Next, we asked whether we could dictate p53 levels by externally applying blue light to the cells expressing PMI_LEXY. We found that cells kept in the dark had ~3-fold higher p53 levels than mock transfected or illuminated cells (Figure 3a,c). Finally, we assessed whether we could also detect higher p21 levels in HCT116 cells kept in the dark. This was indeed the case (Figure 3b,c).

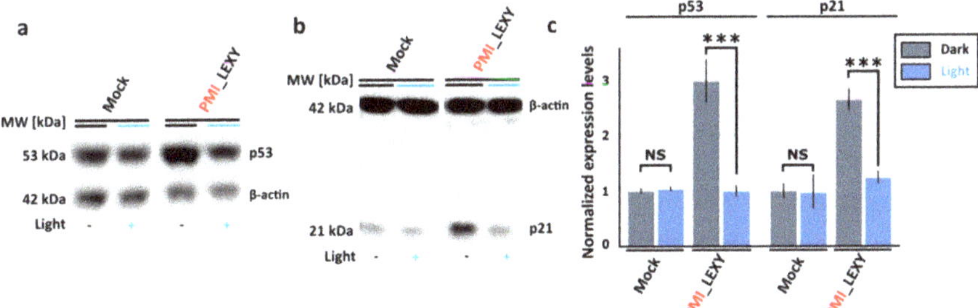

Figure 3. LEXY allows controlling with light endogenous p53 and p21 levels. (a, b) Western blot showing the levels of p53 (**a**) and p21 (**b**) in HCT166 cells. Mock indicates cells transfected with empty pcDNA3.1(+). β-actin was used as loading control. (**c**). Bar plot showing the quantification of three independent western blots with samples as in (a) and (b). Levels are shown normalized to the levels of the mock, dark control. NS $p \geq 0.05$, *** $p < 0.001$, *t*-test.

4. Discussion

p53 protein levels are mainly regulated by the E3 ubiquitin ligase Mdm2, which triggers p53 ubiquitylation and consequent proteasomal degradation. Mdm2 additionally inhibits p53 by binding to its N-terminal transactivation domain (TAD), which is thus not free for contacting the transcription machinery. This second mechanism of p53 inhibition is shared by other proteins, such as MdmX, which do not lead to p53 degradation.

As some tumors are characterized by wild type p53 but aberrant overexpression/activity of one or more of its negative regulators, a promising strategy to combat them is to restore normal p53 signaling by stopping the aberrant inhibition the protein is subject to.

Peptides that bind to Mdm2/MdmX on the same surface used to bind to p53 have been previously developed and shown to be promising.

Here we engineered optogenetic control of the PMI peptide using LEXY, a light-inducible nuclear protein export system previously developed by us. LEXY enables dictating with blue light the localization of any peptide or protein fused to it, in this case PMI and mCherry. This strategy could be applied because we found that PMI needs to be nuclear to function (Figure 1b,c).

It was interesting to see that, in HCT166 cells, nuclearly localized PMI led to higher p21 levels than the control p53-mCherry (with or without additional NLS) (Figure 1b,c). We speculate that this may be due to some interference of the fluorescent protein with p53 transcriptional activity.

It was surprising to see that the MIP peptide, which was reported to elevate p53 levels in HCT116 cells [37], did not have any effect under our experimental conditions (Supplementary Figure S1). As we did not find in the literature other reports confirming the efficacy of this peptide in living cells beyond the work by Shihiedo and colleagues, we reckon further experiments are needed to clarify this issue.

The approach based on LEXY developed here can be used to answer biological questions regarding p53 biology, such as what is the relationship between cellular outcome and p53 levels. The advantage of using light to elevate endogenous p53 levels is that cells with precisely tuned p53 levels can be generated within a single population, making the interpretation of the results easier than when comparing cells cultured separately. Moreover, optogenetics is the best approach to answer questions regarding the effect of the position of cells within a tissue on the cellular outcome, as individual cells can be illuminated. It would be, therefore, possible to study if having higher p53 levels causes different outcomes depending on whether the cells lie in the middle or at the periphery of a tissue.

PMI-LEXY represents an alternative to the existing opto-p53 which we previously established [35]. opto-p53 is the fusion of p53 to LEXY (mCherry was included for visualization purposes). This construct is transcriptionally active and allows controlling p53 target genes by controlling directly the localization of p53, not its levels. In this case, p53-LEXY, that is, an exogenous p53 protein, is the main actuator of the response. In the case of PMI-LEXY, instead, endogenous p53 is the main actuator of the response. The approach based on PMI-LEXY mimics the natural situation, while that based on p53-LEXY suffers from the possibility that alternative pathways are triggered by the high cytoplasmic p53 levels which arise during the illumination phase.

In order to advance this technology towards medical applications, it would be necessary to employ LINuS or other optogenetic tools for which light would lead to p53 accumulation, as the final aim would be to trigger the p53 signaling pathway locally, only in malignant cells. This is not possible with LEXY, as light would have to be constantly applied everywhere else in the body to keep p53 inactive. As LINuS per se works well in HCT116 cells (data not shown), we believe control of PMI with LINuS is achievable with future optimization of the construct. However, blue light penetrates poorly into tissue, thus an approach based on red/infra-red light is highly desirable.

Supplementary Materials: The raw data used to generate the figures in the paper are available online as supplementary figures at www.mdpi.com/xxx/. **Figure S1**: The Mdm2-inhibitory peptide (MIP) does not lead to increased p53 levels in HCT116 cells. (a) Western blot showing the levels of p53 in HCT116 cells transfected with the indicated constructs. Mock indicates cells transfected with empty pcDNA3.1(+). β-actin was used as loading control. (b) Bar plot showing the quantification of three independent western blots with samples as in (a). Data indicate mean ± standard deviation. NS $p >= 0.05$, *t*-test, **Figure S2**: Raw data for the plot shown in Supplementary Figure S1. Only relevant lanes were annotated, **Figure S3**: Raw data for the plot shown in Figure 1c. Only relevant lanes were annotated, **Figure S4**: Raw data for the plot shown in Figure 3c. Only relevant lanes were annotated, **Figure S5**: Raw data for the plot shown in Figure 3c. Only relevant lanes were annotated.

Author Contributions: Conceptualization, P.W. and B.D.V.; Methodology, P.W. and B.D.V.; Software, P.W.; Validation, P.W.; Formal Analysis, P.W. and B.D.V.; Investigation, P.W.; Resources, B.D.V.; Data Curation, P.W.; Writing – Original Draft Preparation, P.W. and B.D.V.; Writing – Review & Editing, B.D.V.; Visualization, P.W.; Supervision, B.D.V.; Project Administration, B.D.V.; Funding Acquisition, B.D.V.

Funding: This work was supported by the BMBF 031L0079 grant to B.D.V.

Acknowledgments: We thank Albert Fabregas for support with the analysis of LINuS variants in HCT116 cells and members of the Di Ventura lab for helpful discussions.

Conflicts of Interest: The authors declare no conflict of interest.

References

1. Belyi, V.A.; Ak, P.; Markert, E.; Wang, H.J.; Hu, W.W.; Puzio-Kuter, A.; Levine, A.J. The origins and evolution of the p53 family of genes. *Csh. Perspect. Biol.* **2010**, *2*, a001198. [CrossRef] [PubMed]
2. Vousden, K.H.; Lane, D.P. p53 in health and disease. *Nat. Rev. Mol. Cell Biol.* **2007**, *8*, 275–283. [CrossRef]
3. Levine, A.J.; Oren, M. The first 30 years of p53: Growing ever more complex. *Nat. Rev. Cancer* **2009**, *9*, 749–758. [CrossRef] [PubMed]
4. Aylon, Y.; Oren, M. The Paradox of p53: What, how, and why? *Cold Spring Harb. Perspect. Med.* **2016**, *6*, a026328. [CrossRef] [PubMed]
5. Hafner, A.; Bulyk, M.L.; Jambhekar, A.; Lahav, G. The multiple mechanisms that regulate p53 activity and cell fate. *Nat. Rev. Mol. Cell Biol.* **2019**, *20*, 199–210. [CrossRef]
6. Kern, S.E.; Kinzler, K.W.; Bruskin, A.; Jarosz, D.; Friedman, P.; Prives, C.; Vogelstein, B. Identification of P53 as a Sequence-Specific DNA-Binding Protein. *Science* **1991**, *252*, 1708–1711. [CrossRef]
7. Chao, C.C. Mechanisms of p53 degradation. *Clin. Chim. Acta* **2015**, *438*, 139–147. [CrossRef]
8. Love, I.M.; Shi, D.; Grossman, S.R. p53 Ubiquitination and proteasomal degradation. *Methods Mol. Biol.* **2013**, *962*, 63–73.
9. Brooks, C.L.; Gu, W. p53 regulation by ubiquitin. *FEBS Lett.* **2011**, *585*, 2803–2809. [CrossRef]
10. Di Ventura, B.; Funaya, C.; Antony, C.; Knop, M.; Serrano, L. Reconstitution of Mdm2-dependent post-translational modifications of p53 in yeast. *PLoS ONE* **2008**, *3*, e1507. [CrossRef]
11. Momand, J.; Zambetti, G.P.; Olson, D.C.; George, D.; Levine, A.J. The mdm-2 oncogene product forms a complex with the p53 protein and inhibits p53-mediated transactivation. *Cell* **1992**, *69*, 1237–1245. [CrossRef]
12. Oliner, J.D.; Pietenpol, J.A.; Thiagalingam, S.; Gyuris, J.; Kinzler, K.W.; Vogelstein, B. Oncoprotein MDM2 conceals the activation domain of tumour suppressor p53. *Nature* **1993**, *362*, 857–860. [CrossRef]
13. Shan, B.; Li, D.W.; Bruschweiler-Li, L.; Bruschweiler, R. Competitive binding between dynamic p53 transactivation subdomains to human MDM2 protein: Implications for regulating the p53.MDM2/MDMX interaction. *J. Biol. Chem.* **2012**, *287*, 30376–30384. [CrossRef] [PubMed]
14. Leslie, P.L.; Ke, H.; Zhang, Y. The MDM2 RING domain and central acidic domain play distinct roles in MDM2 protein homodimerization and MDM2-MDMX protein heterodimerization. *J. Biol. Chem.* **2015**, *290*, 12941–12950. [CrossRef] [PubMed]
15. Capoulade, C.; Bressac-de Paillerets, B.; Lefrere, I.; Ronsin, M.; Feunteun, J.; Tursz, T.; Wiels, J. Overexpression of MDM2, due to enhanced translation, results in inactivation of wild-type p53 in Burkitt's lymphoma cells. *Oncogene* **1998**, *16*, 1603–1610. [CrossRef] [PubMed]
16. Pujals, A.; Favre, L.; Gaulard, P.; Wiels, J. Activation of wild-type p53 by MDM2 inhibitors: A new strategy for lymphoma treatment. *Blood Lymphat. Cancer* **2015**, *5*, 93–100. [CrossRef]
17. Ventura, A.; Kirsch, D.G.; McLaughlin, M.E.; Tuveson, D.A.; Grimm, J.; Lintault, L.; Newman, J.; Reczek, E.E.; Weissleder, R.; Jacks, T. Restoration of p53 function leads to tumour regression in vivo. *Nature* **2007**, *445*, 661–665. [CrossRef]
18. Xue, W.; Zender, L.; Miething, C.; Dickins, R.A.; Hernando, E.; Krizhanovsky, V.; Cordon-Cardo, C.; Lowe, S.W. Senescence and tumour clearance is triggered by p53 restoration in murine liver carcinomas. *Nature* **2007**, *445*, 656–660. [CrossRef]
19. Yu, B.; Tai, H.C.; Xue, W.; Lee, L.J.; Lee, R.J. Receptor-targeted nanocarriers for therapeutic delivery to cancer. *Mol. Membr. Biol.* **2010**, *27*, 286–298. [CrossRef] [PubMed]
20. Linsley, C.S.; Wu, B.M. Recent advances in light-responsive on-demand drug-delivery systems. *Ther. Deliv.* **2017**, *8*, 89–107. [CrossRef] [PubMed]
21. Saneja, A.; Kumar, R.; Arora, D.; Kumar, S.; Panda, A.K.; Jaglan, S. Recent advances in near-infrared light-responsive nanocarriers for cancer therapy. *Drug Discov. Today* **2018**, *23*, 1115–1125. [CrossRef]
22. Goglia, A.G.; Toettcher, J.E. A bright future: Optogenetics to dissect the spatiotemporal control of cell behavior. *Curr. Opin. Chem. Biol.* **2019**, *48*, 106–113. [CrossRef]
23. Ziegler, T.; Moglich, A. Photoreceptor engineering. *Front. Mol. Biosci.* **2015**, *2*, 30. [CrossRef] [PubMed]
24. Halavaty, A.S.; Moffat, K. N- and C-terminal flanking regions modulate light-induced signal transduction in the LOV2 domain of the blue light sensor phototropin 1 from Avena sativa. *Biochemistry* **2007**, *46*, 14001–14009. [CrossRef] [PubMed]

25. Harper, S.M.; Neil, L.C.; Gardner, K.H. Structural basis of a phototropin light switch. *Science* **2003**, *301*, 1541–1544. [CrossRef]
26. Harper, S.M.; Christie, J.M.; Gardner, K.H. Disruption of the LOV-Jalpha helix interaction activates phototropin kinase activity. *Biochemistry* **2004**, *43*, 16184–16192. [CrossRef]
27. Harper, S.M.; Neil, L.C.; Day, I.J.; Hore, P.J.; Gardner, K.H. Conformational changes in a photosensory LOV domain monitored by time-resolved NMR spectroscopy. *J. Am. Chem. Soc.* **2004**, *126*, 3390–3391. [CrossRef]
28. Zayner, J.P.; Antoniou, C.; Sosnick, T.R. The amino-terminal helix modulates light-activated conformational changes in AsLOV2. *J. Mol. Biol.* **2012**, *419*, 61–74. [CrossRef]
29. Kennis, J.; Alexandre, M.; Arents, J.; Van Grondelle, R.; Hellingwerf, K. A base-catalyzed mechanism for dark state recovery in the Avena sativa phototropin-1 LOV2 domain. *Comp. Biochem. Physiol. Mol. Integr. Physiol.* **2007**, *146*, S227. [CrossRef]
30. Lungu, O.I.; Hallett, R.A.; Choi, E.J.; Aiken, M.J.; Hahn, K.M.; Kuhlman, B. Designing photoswitchable peptides using the AsLOV2 domain. *Chem. Biol.* **2012**, *19*, 507–517. [CrossRef] [PubMed]
31. Strickland, D.; Lin, Y.; Wagner, E.; Hope, C.M.; Zayner, J.; Antoniou, C.; Sosnick, T.R.; Weiss, E.L.; Glotzer, M. TULIPs: Tunable, light-controlled interacting protein tags for cell biology. *Nat. Methods* **2012**, *9*, 379–384. [CrossRef]
32. Renicke, C.; Schuster, D.; Usherenko, S.; Essen, L.O.; Taxis, C. A LOV2 domain-based optogenetic tool to control protein degradation and cellular function. *Chem. Biol.* **2013**, *20*, 619–626. [CrossRef]
33. Niopek, D.; Benzinger, D.; Roensch, J.; Draebing, T.; Wehler, P.; Eils, R.; Di Ventura, B. Engineering light-inducible nuclear localization signals for precise spatiotemporal control of protein dynamics in living cells. *Nat. Commun.* **2014**, *5*, 4404. [CrossRef]
34. Yumerefendi, H.; Dickinson, D.J.; Wang, H.; Zimmerman, S.P.; Bear, J.E.; Goldstein, B.; Hahn, K.; Kuhlman, B. Control of protein activity and cell fate specification via light-mediated nuclear translocation. *PLoS ONE* **2015**, *10*, e0128443. [CrossRef]
35. Niopek, D.; Wehler, P.; Roensch, J.; Eils, R.; Di Ventura, B. Optogenetic control of nuclear protein export. *Nat. Commun.* **2016**, *7*, 10624. [CrossRef]
36. Yumerefendi, H.; Lerner, A.M.; Zimmerman, S.P.; Hahn, K.; Bear, J.E.; Strahl, B.D.; Kuhlman, B. Light-induced nuclear export reveals rapid dynamics of epigenetic modifications. *Nat. Chem. Biol.* **2016**, *12*, 399–401. [CrossRef]
37. Shiheido, H.; Takashima, H.; Doi, N.; Yanagawa, H. mRNA display selection of an optimized MDM2-binding peptide that potently inhibits MDM2-p53 interaction. *PLoS ONE* **2011**, *6*, e17898. [CrossRef]
38. Pazgier, M.; Liu, M.; Zou, G.; Yuan, W.; Li, C.; Li, C.; Li, J.; Monbo, J.; Zella, D.; Tarasov, S.G.; et al. Structural basis for high-affinity peptide inhibition of p53 interactions with MDM2 and MDMX. *Proc. Natl. Acad. Sci. USA* **2009**, *106*, 4665–4670. [CrossRef]
39. El-Deiry, W.S.; Tokino, T.; Velculescu, V.E.; Levy, D.B.; Parsons, R.; Trent, J.M.; Lin, D.; Mercer, W.E.; Kinzler, K.W.; Vogelstein, B. WAF1, a potential mediator of p53 tumor suppression. *Cell* **1993**, *75*, 817–825. [CrossRef]
40. El-Deiry, W.S.; Tokino, T.; Waldman, T.; Oliner, J.D.; Velculescu, V.E.; Burrell, M.; Hill, D.E.; Healy, E.; Rees, J.L.; Hamilton, S.R.; et al. Topological control of p21WAF1/CIP1 expression in normal and neoplastic tissues. *Cancer Res.* **1995**, *55*, 2910–2919.
41. Hu, B.; Gilkes, D.M.; Chen, J. Efficient p53 activation and apoptosis by simultaneous disruption of binding to MDM2 and MDMX. *Cancer Res.* **2007**, *67*, 8810–8817. [CrossRef]
42. Saiki, A.Y.; Caenepeel, S.; Cosgrove, E.; Su, C.; Boedigheimer, M.; Oliner, J.D. Identifying the determinants of response to MDM2 inhibition. *Oncotarget* **2015**, *6*, 7701–7712. [CrossRef]
43. Baarlink, C.; Plessner, M.; Sherrard, A.; Morita, K.; Misu, S.; Virant, D.; Kleinschnitz, E.M.; Harniman, R.; Alibhai, D.; Baumeister, S.; et al. A transient pool of nuclear F-actin at mitotic exit controls chromatin organization. *Nat. Cell Biol.* **2017**, *19*, 1389–1399. [CrossRef]
44. Hooikaas, P.J.; Martin, M.; Muhlethaler, T.; Kuijntjes, G.J.; Peeters, C.A.E.; Katrukha, E.A.; Ferrari, L.; Stucchi, R.; Verhagen, D.G.F.; van Riel, W.E.; et al. MAP7 family proteins regulate kinesin-1 recruitment and activation. *J. Cell Biol.* **2019**, *218*, 1298–1318. [CrossRef]

Article

Modular Diversity of the BLUF Proteins and Their Potential for the Development of Diverse Optogenetic Tools

Manish Singh Kaushik [1], Ramandeep Sharma [1], Sindhu Kandoth Veetil [1], Sandeep Kumar Srivastava [2,*] and Suneel Kateriya [1,*]

1 School of Biotechnology, Jawaharlal Nehru University, New Delhi 110067, India; manish13587@gmail.com (M.S.K.); ramandeep31081995@gmail.com (R.S.); kvsindhu@gmail.com (S.K.V.)
2 Department of Biosciences, Manipal University Jaipur, Jaipur 303007, India
* Correspondence: sandeepkumar.srivastava@jaipur.manipal.edu or findsandeep1@gmail.com (S.K.S.); skateriya@jnu.ac.in (S.K.)

Received: 7 April 2019; Accepted: 6 May 2019; Published: 19 September 2019

Abstract: Organisms can respond to varying light conditions using a wide range of sensory photoreceptors. These photoreceptors can be standalone proteins or represent a module in multidomain proteins, where one or more modules sense light as an input signal which is converted into an output response via structural rearrangements in these receptors. The output signals are utilized downstream by effector proteins or multiprotein clusters to modulate their activity, which could further affect specific interactions, gene regulation or enzymatic catalysis. The blue-light using flavin (BLUF) photosensory module is an autonomous unit that is naturally distributed among functionally distinct proteins. In this study, we identified 34 BLUF photoreceptors of prokaryotic and eukaryotic origin from available bioinformatics sequence databases. Interestingly, our analysis shows diverse BLUF-effector arrangements with a functional association that was previously unknown or thought to be rare among the BLUF class of sensory proteins, such as endonucleases, *tet* repressor family (tetR), regulators of G-protein signaling, GAL4 transcription family and several other previously unidentified effectors, such as RhoGEF, Phosphatidyl-Ethanolamine Binding protein (PBP), ankyrin and leucine-rich repeats. Interaction studies and the indexing of BLUF domains further show the diversity of BLUF-effector combinations. These diverse modular architectures highlight how the organism's behaviour, cellular processes, and distinct cellular outputs are regulated by integrating BLUF sensing modules in combination with a plethora of diverse signatures. Our analysis highlights the modular diversity of BLUF containing proteins and opens the possibility of creating a rational design of novel functional chimeras using a BLUF architecture with relevant cellular effectors. Thus, the BLUF domain could be a potential candidate for the development of powerful novel optogenetic tools for its application in modulating diverse cell signaling.

Keywords: photoreceptor; BLUF; modular domain; optogenetics

1. Introduction

Microorganisms respond to changing light conditions using an evolved repertoire of photoreceptors that perceive light and execute a light-dependent control of regulatory 'output' domains [1]. Blue-light using flavin (BLUF) protein photoreceptors respond to blue light and are often coupled with different effector domains (enzymes or transcriptional regulators) to generate the full range of combinations to regulate photo-adaptive responses [2–5]. However, proteins having the BLUF domain with an extended C-terminus only have also been reported, and their responses were controlled by light-dependent protein-protein interactions [6–9]. Upon illumination, the isoalloxazine moiety of the BLUF domain associated flavin chromophore (flavin adenine dinucleotide; FAD, or flavin adenine mononucleotide; FMN, or riboflavin; RF) absorbs blue light [10], and undergoes structural rearrangements to modulate the communion between BLUF and the effector domains [11]. Unlike the complex mechanisms of photo-transformation in other photoreceptors, the BLUF domain, upon illumination, mainly shows a hydrogen bond rearrangement around the flavin cofactor, which causes a 10–15 nm red shift in the BLUF absorbance peak [11]. The photo-activation of the BLUF domain is due to the involvement of a conserved glutamine and tyrosine residues [12–14]. The hydrogen bond rearrangement involves a unique ability of the BLUF domain, i.e., photo-induced proton-coupled electron transfer (PCET), which enables them to switch between receptor and signaling states [15–17]. The photo-activation and structural rearrangements around the chromophore of the BLUF domain are transmitted as a signal for the activation of an associated effector domain. The BLUF domains are considered as an attractive model to investigate new paradigms of photo-induced signaling. BLUF domains have a modular architecture; hence, they may be functionally fused to different effector domains, as observed for the modular light oxygen voltage (LOV) photoreceptors [18–25]. Barends and coworkers have characterized a full-length active photoreceptor, BlrP1, from *Klebsiella pneumoniae*, which is composed of BLUF and EAL (Glutamine-Alanine-Leucine) as the sensor/output domain combination, respectively [4]. The EAL domain is a conserved signature motif which hydrolyses cyclic dimeric GMP (c-di-GMP) and is involved in the regulation of motility, biofilm formation, virulence and antibiotic resistance in the bacteria [26–29]. When exposed to light, the BLUF domain from BlrP1 activates the EAL domain via an allosteric communication relayed through conserved domain-domain interfaces [4]. In *Escherichia coli*, YcgF is another photoactivated protein with the BLUF-EAL domain that has been reported [30]. However, unlike other EAL domain proteins, YcgF acts as a transcriptional regulator and controls the YcgF/YcgE pathway, which regulates the synthesis of small regulatory proteins. These small regulatory proteins are involved in the modulation of biofilm functions via the Rcs two-component pathway, necessary for *E. coli* to sustain the adverse environment [30]. In many bacteria, proteins with tandem GGDEF (diguanylate cyclase; DGC)/EAL (phosphodiesterase) domains were also reported to be involved in the c-di-GMP turnover, which modulates a variety of functions ranging from the functional modification of cell surface components, the expression of extracellular signaling molecules, virulence and motility [31,32]. Photoactivated BLUF associated adenylyl cyclase homology domains (CHD) were also investigated and well characterized in several microorganisms, where they are specifically involved in the catalytic conversion of ATP to cyclic AMP (cAMP), which regulates the downstream signal transduction [33–38]. PAS domains (Per/ARNT/Sim) are one of the broadly spread domains involved in sensing variations in light, oxygen, redox potential and the binding of small ligands [39]. The role of PAS domains is diverse, and few reports demonstrated the involvement of PAS in domain chromophore attachment [40], light-regulated protein-protein interactions [41] and in complementary chromatic adaptation (CCA) [42].

Optogenetics is a recently developed molecular tool that combines genetic and optical methods and enables us to modulate specific functions in any cell or tissue using light in a controlled manner [43]. Microbial, algal opsins and natural light regulated ion channels have been reported to be versatile and good actuators for optogenetic applications [44]. However, BLUF domains, due to their small size, solubility, reversibility, temporal precision and diverse association with a wide variety of effectors, could be engineered for the light-dependent modulation of a wide range of cellular signaling [44]. Prominent examples include BLUF containing photoactivated adenylyl cyclase from *Beggiatoa* sp. (bPAC) efficiently activating cyclic-nucleotide-gated ion channels in neurons. The mutagenic variant of bPAC, BlaG, yielded a higher level of light-induced production of cGMP than cAMP [11,45]. In the present study, we have characterized the modular diversity of the BLUF domain coupled proteins that could be valuable in the development of novel synthetic photoswitches and we expand the scope of the optogenetics modulation of novel cellular signaling within a functional expression in the appropriate living system.

2. Materials and Methods

2.1. Database of Sequences used in this Analysis

The BLUF domain encoding protein sequences were retrieved from the National Center for Biotechnology Information (NCBI; https://www.ncbi.nlm.nih.gov/), and each of them was subjected to a conserved domain search using the Conserved Domain Architecture Retrieval Tool (CDART; https://www.ncbi.nlm.nih.gov/Structure/lexington/lexington.cgi) [46]. The 34 uncharacterized BLUF domain containing proteins were selected for a further analysis. For each protein, sequences encoding the BLUF domain were selected and aligned for the homology analysis using the BioEdit tool [47]. The Multiple EM for the Motif Elicitation (MEME) suite (http://meme-suite.org) was employed to scan conserved motifs throughout the sequences [48]. The interacting partners for each of the output domains were predicted using the String version 11 [49].

2.2. Phylogenetic Analysis

A phylogenetic analysis involving 34 BLUF sequences from different organisms was performed by employing the Maximum Likelihood method, based on the JTT matrix-based model [50]. Gaps were eliminated from the sequences. The tree with the highest log likelihood (−955.3175) is shown. The values shown with the branches represent the percentage of trees clustering with the associated taxa. The JTT matrix of pairwise distance was subjected to Neighbor-Join and BioNJ algorithms for the construction of the first tree(s), which was then used to select the topology with a higher log likelihood value. Phylogenetic analyses were performed using MEGA6 [51].

2.3. Analysis and Homology Modeling of the BLUF Domain

The annotated sequences were further analyzed for the putative secondary structures and function using Predict Protein (https://www.predictprotein.org/). Based on the secondary structure analysis, a two-dimensional topology was generated using POTTER (http://wlab.ethz.ch/protter/#) [52]. Three-dimensional models of the predicted BLUF sequences and associated effector domains were created using the Phyre2 modeling tool [53], employing an integrated combinatorial approach comprising comparative modeling, threading, and *ab initio* modeling [54]. All the energy-minimized models of the annotated BLUF domains and BLUF, in combination with the effector domains, were further evaluated for structural errors and the stereochemistry quality, as well as for manual curation. The variety of the models were performed, in terms of bond angles, distances, stereochemical analysis, and vocabulary, by the UCLA Structure Analysis and Verification Server (SAVES) with the PROCHECK and ERRAT programs [55,56]. Finally, the most acceptable models were finalized based on Ramachandran plot analysis and structural fit for each annotated sequence. All of the predicted BLUF sequences were aligned for an analysis of conservation and variation of residues using the Clustal Omega program.

3. Results and Discussion

3.1. BLUF Sequences, Modular Domains and Phylogenetic Analysis

We assessed the residues that are conserved and important for the substrate specificity in the respective orthologs. The details for each protein sequence and domain architecture are given in Table 1 and Figure 1.

Table 1. Blue light using flavin (BLUF) modular domains from different organisms.

Domains	Accession No.	BLUF Length	Score	Organisms	Probable Modulations
BLUF + EAL	ARH96915.1	2–91	e^{-28}	*Escherichia coli*	Regulate diguanylate cyclases and Phosphodiesterase activity
BLUF + PsiE	EGW22399.1	2–92	e^{-36}	*Methylobacter tundripaludum* SV96	NP*
BLUF + CHD	EHQ08139.1	3–93	e^{-28}	*Leptonema illini* DSM 21528	Regulate adenylyl and guanylyl cyclase activity
BLUF + B_{12} Binding domain	ABP71929.1	17–106	e^{-35}	*Rhodobacter sphaeroides* ATCC 17025	Broadens BLUF photosensing ability
BLUF + PRK09039 superfamily	WP_012321331.1	5–99	e^{-36}	*Methylobacterium radiotolerans*	Regulate phosphoribulokinase, uridine kinase and pantothenate kinase activity
BLUF + DNA_pol3_gamma3 superfamily	AIQ92835.1	5–99	e^{-36}	*Methylobacterium oryzae* CBMB 20	Regulate DNA replication process
BLUF + REC	AMR27912.1	150–240	e^{-37}	*Hymenobacter* sp. PAMC 26554	Regulate bacterial chemotaxis
BLUF + P450 superfamily	WP_045444510.1	15–108	e^{-35}	*Psychrobacter* sp. P11F6	Regulate oxidative degradation of steroids, fatty acid and xenobiotics
BLUF + AcrR	WP_058726129.1	3–94	e^{-28}	*Curtobacterium luteum*	Regulate antibiotic resistance in bacteria
BLUF + TetR_C_6	WP_051596720.1	4–92	e^{-34}	*Curtobacterium sp.* UNCCL17	Regulate antibiotic resistance in bacteria
		343–417	e^{-19}		
BLUF +Endonuclease_NS	WP_058743091.1	17–83	e^{-30}	*Drosophila eracta*	Modulate hydrolase activity, nucleic acid and metal ion binding
BLUF + PAS	WP_058511962.1	4–88	e^{-13}	*Legionella steelei*	Regulate cellular signaling processes
BLUF + AraC	WP_053973760.1	228–318	e^{-19}	*Polaribacter dokdonensis*	Control synthesis of structural components of arabinose metabolism
BLUF + Abhydrolase super family	XP_008692928.1	58–99	e^{-15}	*Ursus maritimus*	Modulate hydrolytic enzyme activity
BLUF + ENDO3c Superfamily	EYD78138.1	1–89	e^{-27}	*Rubellimicrobium mesophilum* DSM 19309	Control DNA repair regulation
BLUF + ANK	EJY80769.1	547–593	e^{-13}	*Oxytricha trifallax*	Modulate protein-protein interaction
BLUF + DUF1115 Superfamily	WP_0229622806.1	5–101	e^{-37}	*Pseudomonas pelagia*	NP*
BLUF + RhoGEFSuperfamily	XP_025342216.1	171–262	e^{-38}	*Pseudomicrostroma glucosiphilum*	Control activation of Rho family GTPases
BLUF + PDZ	jgi_Bigna1_85551	55–144	e^{-17}	*Bigelowiella natans* CCMP 2755	Regulate membrane-bound cell signaling
BLUF + AANH_like Superfamily	Jgi_Schag1_101311	273–366	e^{-24}	*Schizochytrium aggregatum* ATCC 28209	NP*
BLUF + EAL + GGDEF	AFL74487.1	456–546	e^{-33}	*Thiocystis violascens* DSM 198	Regulation of c-di-GMP level
BLUF + EAL+ PRK15043 superfamily	CDW60191.1	2–62	e^{-25}	*Trichuris trichuris*	NP*
BLUF + GGDEF+ PAS	WP_058465269.1	1–91	e^{-25}	*Legionella cincinnatiensis*	NP*
BLUF + GGDEF+ COG5001 Superfamily + PAS	WP_058516015.1	3–85	e^{-23}	*Legionella santicrucis*	NP*
BLUF + COG5001 + PBP1_NHase + PAS	ADC61983.1	974–1064	e^{-38}	*Allochromatium vinosum* DSM 180	NP*

Table 1. *Cont.*

BLUF + PsiE +BaeS superfamily	WP_014148160.1	3–93	e^{-42}	*Methylomicrobium alcaliphilum*	NP*
BLUF + CHD + LRR_R1 Superfamily	Q8S9F2.1	57–136	e^{-24}	*Euglena gracilis*	NP*
		468–554	e^{-27}		
BLUF + CHD + Med 26_M Superfamily	XP_013758351.1	98–186	e^{-19}	*Thecamonas trahens* ATCC 50062	NP*
		673–756	e^{-18}		
BLUF + PRK11633 + DNA pol3 gamma3 family	WP_048452447.1	5–99	e^{-42}	*Methylobacterium tarhaniae*	NP*
BLUF + TetR_C_11 family + Fer2_2 superfamily	WP_058743091.1	4–92	e^{-42}	*Curtobacterium citreum*	NP*
BLUF + GAL4 + Fungal TF MHR	ORY86082.1	10–100	e^{-38}	*Protomyces inouyei*	Regulate galactose induced genes
BLUF + SRPBCC Superfamily + RGS Superfamily	BAV14116.1	518–612	e^{-26}	*Naegleria fowleri*	NP*
BLUF + SRPBCC + BTB + DUF35522 Superfamily	EFC49155.1	1132–1226	e^{-28}	*Naegleria gruberi*	NP*
BLUF + SRPBCC + FH2 + Drf_FH1 + PRK13729 + SMC_N	XP_002669619.1	1971–2050	e^{-28}	*Naegleria gruberi*	NP*

BLUF- Blue light using flavin; EAL- Glutamine/Alanine/Leucine; PRK- Phosphoribulokinase; PAS- Per/Arnt/Sim; PBP- Phosphatidylethanolamine-Binding Protein; CHD- Cyclase homology domain; LRR- Leucine-rich repeats; Med26- Mediator of RNA polymerase II transcription subunit 26; B_{12}- Vitamin B_{12}; DNA pol-DNA polymerase; REC- cheY-homologous receiver domain; HTH-Helix turn helix; Endo3- Endonuclease 3; ANK- Ankyrin repeats; RGS- Regulator of G protein signaling; DUF- Domain of unknown function; RhoGEF- Guanine nucleotide exchange factor for Rho/Rac/Cdc42-like GTPases; GAL4- GAL4-like Zn(II)2 Cys6 (or C6 zinc) binuclear cluster DNA-binding domain; MHR- Middle homology region; PDZ- PSD95/Dlg1/zo-1; SRPBCC- START/RHO_alpha_C/PITP/Bet_v1/CoxG/CalCligand-binding; BTB- Broad-Complex, Tramtrack and Bric a brac; FH2- Formin Homology 2; Drf- Diaphanous related formins; SMC_N- N terminus of structural maintenance of chromosomes. Sequences WP_051596720.1 from *Curtobacterium* sp. UNCCL17; Q8S9F2.1 from *Euglena gracilis;* and XP_013758351.1 from *Thecamonas trahens* ATCC 50062 are strongly predicted to contain two BLUF domains on either side of their respective effector domains. NP*: Probable modulated function cannot be predicted, but requires a detailed study.

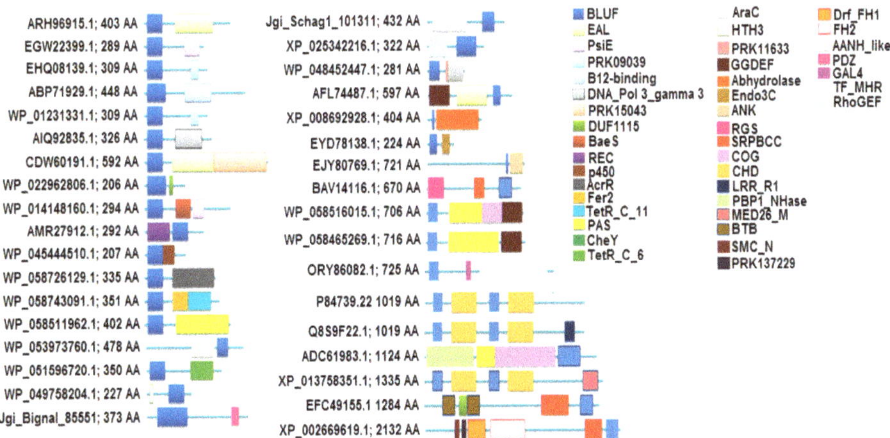

Figure 1. Schematic representation of the different blue light using flavin (BLUF) modular domain containing proteins. The accession numbers were taken from National Center for Biotechnology Information (NCBI). The mentioned "AA" indicates the amino acid numbers of the particular BLUF modular protein.

The analysis of 34 BLUF sequences revealed that residues forming BLUF catalytic core are conserved throughout for interaction with the flavin chromophore (Figure 2). The amino acids tyrosine (Y), asparagine (N), glutamine (Q), and tryptophan (W) or methionine (M), which are crucial for the photodynamics and photocycle of the BLUF domain, are highly conserved (Figure 2). Our analysis further confirms that tyrosine, glutamine, and tryptophan (or methionine) are critical for the substrate specificity of the BLUF sequences, as has been reported earlier [6,11,33,57]. The photo-activation of the BLUF domains actually involves conserved glutamine and tyrosine residues, where glutamine contributes to hydrogen bond formation in a dark state [11–14]. However, in light adapted conditions, a hydrogen bond rearrangement (tautomerization) occurs to form a new hydrogen bond with tyrosine [11]. A motif analysis revealed three different conserved motifs (Figure 2) among different BLUF sequences. Each motif has its importance, as each of them was comprised of an essential amino acid residue involved in the regulation of the BLUF photocycle and photodynamics [6,11].

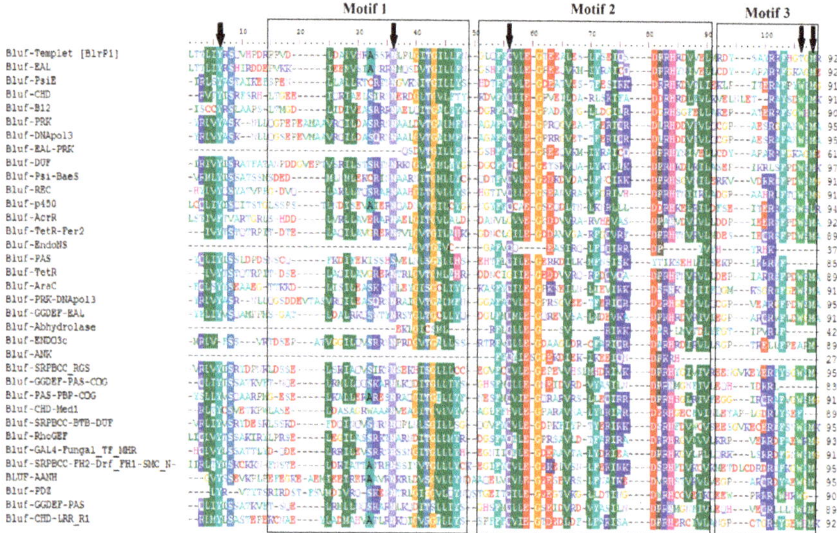

Figure 2. Multiple sequence alignment of the different BLUF modular domains depicting conserved amino acids. The black arrow indicates conserved amino acids crucial for regulating the flavin binding pocket, photocycle and photodynamics of the BLUF domain containing proteins [6,11]. The sequences under the solid boxes represent the conserved motifs of the BLUF domain. The conserved motifs were predicted using the Multiple EM for the Motif Elicitation (MEME) suite.

The phylogenetic analysis of the selected sequences of 34 BLUF domains divides them into four distinct clusters (I–IV), which seem to differ from each other mainly by the presence of eukaryotic and prokaryotic BLUF proteins in them (Figure 3). As observed from the phylogenetic analysis, the BLUF domain sequences of eukaryotic and prokaryotic origins are evolutionarily intermixed. Each cluster (except cluster IV) was further subdivided into sub-clusters, which represent closely related BLUF domain sequences broadly associated with similarkind of effector domains (Figure 2).

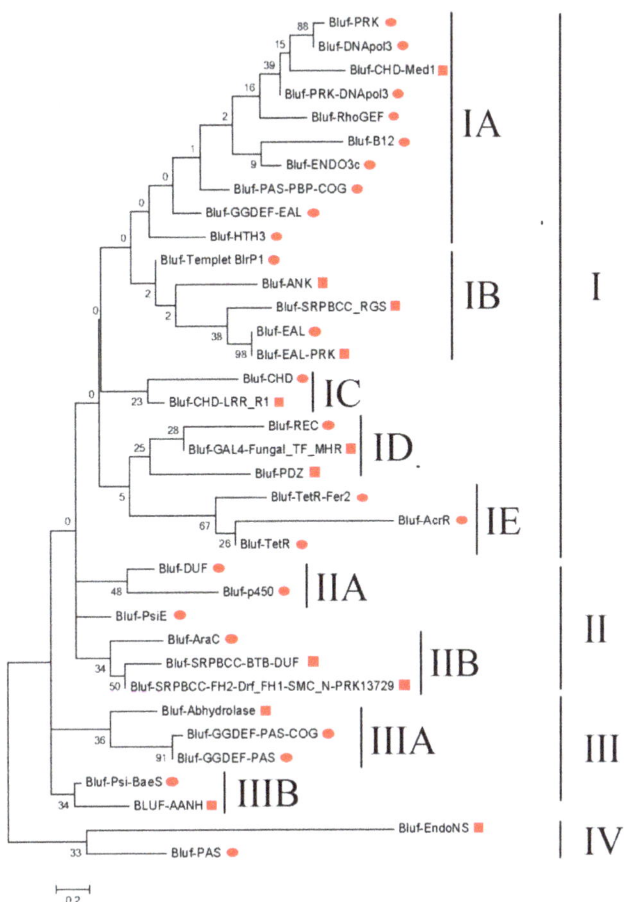

Figure 3. Phylogenetic analysis by the Maximum Likelihood (ML) method. The analysis was done using 34 amino acid sequences of the modular BLUF domain containing proteins. All positions containing gaps and missing data were eliminated. There were a total of 17 positions in the final dataset. A solid red circle represents protein sequences of prokaryotic origin, and a solid red square represents protein sequences of eukaryotic origin. Evolutionary analyses were conducted using MEGA6 [51].

3.2. Modular Diversity of BLUF Domains

In the present communication, we have selected 34 different modular architectures for the BLUF domain in combinations with various effector domains (Table 1; Figure 1). The annotated BLUF domains in our database sequences range from 80 to 100 amino acids, situated mostly at the N-terminal region of the effector domains. However, the predicted BLUF sequences in our database show a high sequence similarity across the BLUF region, with E-values in the range of e^{-40} to e^{-120}. The residues show an upper conservation pattern around the flavin binding region (Figure 4). The secondary structural analysis indicates that the BLUF sequences are composed mostly of five β strands and two α helices. The E-value, sequence length and sequence conservation details are summarized in Table 1. Three-dimensional models of the BLUF domains were constructed using the Phyre2 modeling software [53]. The BLUF domains reveal conserved βαββαββ fold conformations in which two α-helices surround a five-stranded antiparallel β-sheet platform (Figure 4).

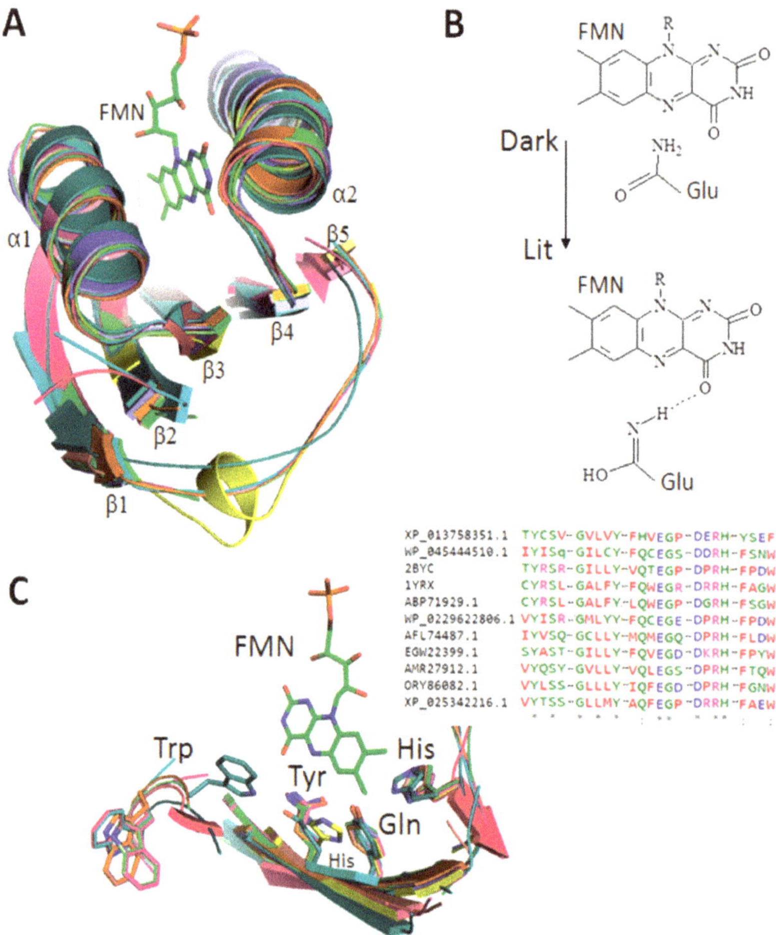

Figure 4. Models of the BLUF domains and details of its flavin binding pocket. (**A**) Superposition of the modelled BLUF domains using the Phyre server with a crystal structure of *Rhodobacter sphaeroides* BlrB (PDB: 2BYC) (magenta). The homology models include the annotated BLUF domains from WP_045444510.1 (*Psychrobacter sp.*) (Green tone); WP_014148160.1 (*Methylomicrobium alcaliphilum*) (Brown tone); XP_025342216.1 (*Pseudomicrostroma glucosiphilum*) (Blue tone); AFL74487.1 (*Thiocystis violascens*) (Violet tone); EHQ08139.1 (*Leptonema illini*) (Yellow tone); ABP71929.1 *Rhodobacter sphaeroides* ATCC 17025 (Pink tone); WP_0229622806.1 (*Pseudomonas pelagia*) (Skyblue tone); and ORY86082.1 (*Protomyces inouyei*) (Smudge tone). The domain boundaries of the modelled BLUFs are mentioned in Table 1. (**B**) The BLUF photocycle scheme shows the protein/FAD interactions through the hydrogen bonding pattern of the flavin moiety with conserved glutamine upon illumination. (**C**) Superposition of the flavin binding pocket in the BLUF models in comparison to *Rhodobacter sphaeroides* BlrB (PDB: 2BYC) and the BLUF domain of AppA (PDB: 1YRX). The side chains of residues with potentially important roles in catalysis and/or substrate binding are shown as stick models and are labelled. The selected regions of the same are shown in a reduced multiple sequence alignment.

We modeled the binding modes of the flavin chromophore based on the observations in the known BLUF domain structures [33,57]. The isoalloxazine ring of flavin can be readily accommodated in the pocket in the models sandwiched between two α helices, suggesting that the active site is appropriately formed in BLUF models (Figures 4 and 5). Spectroscopic analyses and photochemistry of BLUF proteins have shown that flavin chromophore facilitates blue light-induced electron transfers by a hydrogen bond rearrangement between flavin N5 and O4 and conserved tyrosine, glutamine, and tryptophan or methionine of the BLUF [14,17], leading to a spectral shift from blue to red and the transmission of the signal further downstream. Sequence alignment studies of 34 BLUF domain containing sequences, in this study and elsewhere, show the conservation of tyrosine and glutamine residues in the β1 and β3 strands of the BLUF fold (Figure 4), and this has been reported to be critical for the photochemical reaction, as mutations in these amino acids result in the loss of its ability to perceive light [12–14,17,58]. In our sequence database, BLUF photo domains are fused at the N-terminus of a wide array of different effector domain modules, viz. kinases, phosphatases, phosphodiesterases, anti-sigma factors, DNA binding domains, a transcriptional repressor (TetR), PAS, endonuclease and PBP proteins. Some of these have been modeled with >90% confidence using the Phyre2 modeling program and are presented in Figure 4. These domains are part of various photoreceptors [1,59,60]. Our structural and field mapping analysis also shows that sequences like TetR_C_6, SRPBCC1 superfamily, GGDEF1 COG5001 superfamily, COG5001 super family PBP1_NHase PAS superfamily, and CHD4 Med26_M super family comprise of two BLUF domains at each end with an effector domain sandwiched in-between (Table 1). LOV photoreceptors have also been reported to be present in two copies, along with effector sequences, but both are situated at the N-terminus of effector domains [61]. It remains to be seen if the role of two BLUF domains on either side of the effector is to diversify the signal further or to play the role of a signal transduction further downstream with the help of the auxiliary linker helix found toward the C-terminus of the conserved BLUF domain in our database. The function of these short auxillary helical stretches, located in the association of several BLUF and LOV photoreceptors, is not known precisely, but it seems likely that they mediate the signal progression between their photosensors and the effector domain, as has been predicted in earlier reports [62].

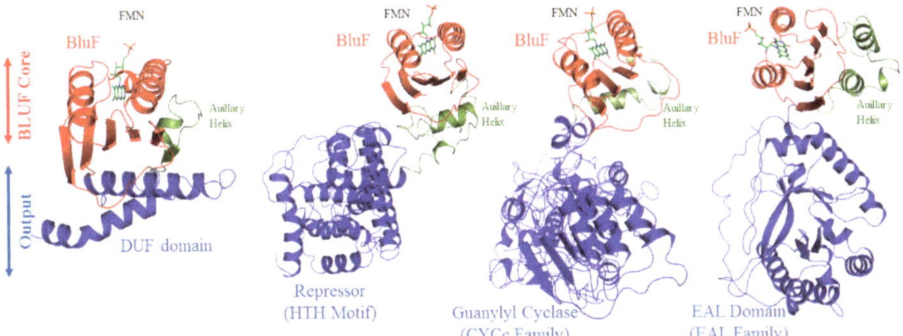

Figure 5. Representation of 3D structural models of BLUF (Red) and effector domain combinations (Blue) using the Phyre server [54]. (**A**) WP_0229622806.1 (*Pseudomonas pelagia*) (206 aa) codes a combination of the BLUF (5–101) and DUF domain (99–181). The model was generated with 100% confidence covering 1–186 residues using the *R. sphaeroides* AppA (PDB: 4HH0) as the template. (**B**) WP_051596720.1 (*Curtobacterium sp.* UNCCL17) (470 aa) codes for BLUF in combination with the transcriptional repressor (224–327). The model was generated with >90% confidence covering 1–350 residues using 12 different templates. (**C**) Q8S9F2.1 (*Euglena gracilis*) (1019 aa) codes for BLUF in combination with cyclase homology domains (CHDs), which are part of the class III nucleotydylcyclases (20–379). The model was generated with >90% confidence covering 1–800 residues using 13 different templates. (**D**) ARH96915.1 (*Escherichia coli*) (403 aa) codes for BLUF (2–93) in combination with the EAL signaling domain (150–389). The model was generated with 100% confidence covering 1–389 residues using *K. pneumoniae* BlrP1 (PDB: 3GFZ) as the template. The structures are represented as interactive coloured ribbons. The model images were generated using PyMol (http://www.pymol.org) [63].

3.3. BLUF Modules in Association with the Effector Domains

The occurrence of the BLUF domain in different associations reveals the diversity and abundance of this modular domain in a wide range of organisms. Our sequence and fold analyses confirm various types of effector domains fused with the BLUF domains (Figure 5).

3.3.1. EAL and GGDEF Domain

The EAL and GGDEF domain-containing proteins are widely distributed among bacteria and are involved in the regulation of the cellular level of the universal signaling molecule bis-(3′,5′)-cyclic-guanosine monophosphate (c-di-GMP), where the former act as diguanylate cyclases (DGCs) while the latter ones are phosphodiesterases (PDEs) [64]. The c-di-GMP generally controls a variety of signaling pathways associated with cell differentiation, bacterial adhesion and biofilm formation, bacterial motility, the colonization of host tissues and virulence [65,66]. In the EAL-GGDEF domain containing proteins, the GGDEF and EAL motifs in the active sites are crucial for the DGC and PDE enzyme activities [67–69]. Although most of the proteins involved in c-di-GMP signaling contain the GGDEF/EAL domains as a single polypeptide, possessing both the DGC and PDE enzyme activities, in some cases the GGDEF and EAL domain is also found alone [68]. In some of the EAL-GGDEF domains containing proteins, it was also reported that one [70,71] or both [72,73] of these domains are catalytically inactive due to a lack of respective GGDEF and EAL motifs that are crucial for the enzymatic function. In these proteins, the inactive domains either act as regulators [74] or c-di-GMP effectors [72,73]. Yang and coworker have reported a FimX-like protein (Flip) with a degenerate EAL-GGDEF domain which interacts with the PilZ-Domain protein to control virulence in *Xanthomonas oryzae pv. oryzae* [75]. YhdA is another protein with a degenerate EAL-GGDEF domain that promotes the turnover of CsrB and CsrC (small RNAs), which reduce the expression of the flhDC (flagellar master regulator) by sequestering the CsrA (RNA-binding protein) [76]. Several EAL-GGDEF domains containing proteins are also reported to have N-terminal sensory domains that can regulate the GGDEF and/or EAL domain functions. In the present study, we have selected an EAL only and EAL-GGDEF domain-containing proteins with the blue light using flavin (BLUF) domain as a sensory domain from *E. coli* and *Thiocystis violascence*, respectively (Table 1).

Barends and coworker have characterized the *Klebsiella pneumoniae* BlrP1 photoreceptor protein biochemically, structurally, mechanistically, and have elucidated the mechanism of the light-induced regulation of the EAL domain (PDE) via the BLUF sensor domain [4]. Upon light illumination, the structural change in the flavin binding pocket (Trp replace Met) of the BLUF domain cross-activates the EAL domain via allosteric communication and increases the PDE activity [4]. The photo-dependent alteration of the BLUF–EAL interactions influences the quaternary structure, the EAL–EAL interface at the dimerization helix, the compound helix α5EAL and the loop connecting it to β5EAL, which are implicated in the EAL activation [4,77]. In the BlrP1 photoreceptor, the BLUF domain shares a similar architecture, as shown by other BLUF domains from different organisms, where the central BLUF domain (N-terminus) is surrounded by two helices (helical cap; C-terminus) [78–80]. However, an additional EAL output domain has been identified in the BlrP1 protein, which is connected to the BLUF domain via a 50 Å long linker peptide (triosephosphate isomerase (TIM)-barrel fold) [4]. The EAL active site in the BlrP1 photoreceptor involved Glu188, Asn239, Glu272, Asp302, Asp303, Lys323 and Glu359 [4]. We used this well-characterized BlrP1 as a template and aligned it pairwise with EAL only (*E. coli*), EAL-GGDEF (*Thiocystis violascence* DSM 198), and EAL-PRK15043 domains (*Trichuris trichuris*), and observed that most of the amino acids constituting the EAL active site in BlrP1 are highly conserved through all of the EAL domain-containing proteins (Figure S1), which suggested that the EAL domain-containing proteins possess a similar mechanism for the PDE activity.

The query protein sequence (ARH96915.1) containing the EAL output domain was subjected to a protein-protein interaction analysis, which revealed several interacting partners involved in the regulation of different signaling pathways (Figure S7a; Table S1). The protein-protein interaction analysis showed that most of the interacting partners for this photoreceptor are either EAL domain-containing

PDEs (JD73_03740) or GGDEF domain-containing DGCs (JD73_23675, JD73_25605, JD73_23680, YeaP, and YdaM). However, the interactions with YcgZ (two-component connector protein) and YcgE (Mer-like repressor protein/transcriptional regulator) have also been revealed, indicating their involvement in the regulation of bacterial biofilm formation [30]. Tschowri et al. [30] have characterized the previously unknown function of the BLUF-EAL domain-containing protein, YcgF, from *E. coli*, and suggested that upon blue light irradiation, this protein acts like an antirepressor. The antirepressor YcgF removesYcgE (Mer-like repressor) from the promoter, and resumes the expression of different small regulatory proteins (YmgA and YmgB). These small regulatory proteins (YmgA and YmgB) utilize the RcsC/RcsD/RcsB two-component phosphorelay system to activate the production of colanic acid, a biofilm matrix component, and to decrease adhesive curli fimbriae [30]. The query protein may also interact with AriR/YmgB (regulator of acid resistance influenced by indole), another biofilm-related protein involved in the regulation of acid resistance in *E. coli* [81]. A string analysis also revealed the possible interaction between the query protein and regulatory protein, LuxR, which regulates quorum sensing in the bacterial system [82,83].

3.3.2. PsiE Domain

In *E. coli*, the phosphate-starvation-inducible (*psiE*) gene is positively and negatively regulated by both PhoB and cAMP-CRP (cAMP receptor protein), which are respectively involved in the phosphate and carbon metabolism [84]. The phosphate and carbon sources regulate the *psiE* gene by using the *lacZ* and *chloramphenicol acetyltransferase* gene (*cat*) fusions, respectively [84]. Although the function of PsiE has not yet been determined, sometimes it has been predicted to have features like DNA-binding protein inhibitor-related, putative transcriptional regulators or hypothetical DNA binding proteins (IPR020948).

3.3.3. Cyclase Homology Domain (CHD)

The cyclase homology domains (CHDs) are the catalytic domains of eukaryotic and prokaryotic nucleotidylcyclases, i.e., adenylyl cyclases (ACs) and guanylyl cyclases (GCs), which belong to the evolutionary diverse class III nucleotidylcyclases. CHDs are reported as three different structural forms, i.e., heterodimers (mammalian CHDs), pseudoheterodimers (metazoan CHDs) and homodimers (bacterial and protozoan CHDs) [85]. Heterodimeric and pseudoheterodimeric CHDs have a single catalytic pocket sharing catalytic amino acid residues at the dimer interface, while homodimeric CHDs have two separate catalytic pockets, each of which contributes the CHD determinant [85]. In spite of having two potential catalytic pockets, several enzymes with homodimeric CHDs (for example, eukaryotic class III nucleotydylcyclases) may have only one catalytically competent site [86]. Although CHDs are structurally diverse, all of them have a conserved structural component (i.e., a helical region) mutating which compromises the stability and active dimeric conformation of a protein [87]. All CHDs have a common catalytic mechanism in which they require two magnesium or manganese ions to bind a polyphosphate group of the nucleotide, followed by nucleophile activation.

Most CHDs, except a few (Rv1359 from *M. tuberculosis*), are reported to exist in combination with different regulatory modules, thus making it possible to perceive the variety of signals and the regulation of the intracellular cAMP generation [85]. In the present communication, we have selected three such proteins, with the BLUF domain in combination with the CHD domain, having the accession numbers EHQ08139.1 (from *Leptonema illini* DSM 21528), Q8S9F2.1 (from *Euglena gracilis*) and XP_013758351.1 (from *Thecamonas trahens* ATCC 50062) (Table 1). In *Euglena gracilis*, CHDs (adenylyl cyclases) occur in combination with the BLUF domain, with an overall domain arrangement of BLUF1CHD1BLUF2CHD2, where the flavin chromophore senses blue light and stimulates the adenylyl cyclase activity [35]. Furthermore, we have performed a alignment of sequences representing the CHD domain from selected proteins against the progression of the well-characterized template CHD domain of bPAC (Figure S2). The multiple sequence alignment revealed that active site residues involved in the nucleotide binding (i.e., Asn257-His266, Lys263-Met264 (forming the β4AC- β5AC tongue), Gly259-Asn178 (forming the α2AC helix) and Lys263–Thr196, Asp265–Phe198, and His266–Lys197 (forming the β2AC- β3AC hairpin) [33]) are highly conserved among all three selected BLUF-CHD domain-containing proteins (Figure S2). In PACs, under dark conditions, the orientation and arrangement of an individual amino acid in the active site (i.e., Thr267 and Lys197 (bound to Phe198)) render the conformation of AC inactive. On the other hand, upon illumination, the change in the interaction between Asn25 and His266 resulted inthe correct orientation of Thr267 required for the communication with the adenine base; additionally, the Lys197 is detached from Phe198, thus providing space for the adenine base to enter the active site more deeply [33]. We also performed a protein-protein interaction analysis, which revealed the possible interacting partners for CHD domain-containing proteins, which range from phosphodiesterases, the RNA polymerase subunit β, and the DNA helicase to another adenylate cyclase/guanylate cyclase associated with the GAF and PAS/PAC sensor (Figure S7b and Table S1).

3.3.4. PAS Domain

PAS (Per-Arnt-Sim) domain-containing proteins are widely distributed among all domains of life. PAS domain acts as a sensor, generally found at the N terminus of sensory and signaling transduction related proteins, and detect a variety of stimuli and regulating the functions of a diverse array of effector domains [88,89]. Members of the PAS domain family can bind a diverse range of small-molecule metabolites [22], which could either directly act as a signal and be involved in initiating a cellular signaling response [90], or which could serve as a cofactor and respond to subsequent messages like gas molecules, redox potential, or photons [39]. Although PAS domain-containing proteins are chemically and functionally diverse, almost all PAS domains have a conserved core comprised of a five-stranded antiparallel β-sheet and several α-helices, which are responsible for the generation and propagation of a signal to the adjoining effector domain. In the present study, we have selected four proteins, three of them from different *Legionella* strains and one from *Allochromatium vinosum* DSM 180, respectively, having combinations of the BLUF and PAS sensor domain (Table 1). We performed the alignment of sequences representing the PAS domain in four different proteins against a well-characterized photoactivated yellow protein (PYP; 1NW_Z) from *Halorhodospira halophila* [91] (Figure S3). As discussed earlier, PAS domains are structurally diverse; the same is revealed from the multiple sequence alignment analysis. Although, the PAS core (5′ NAAEGDIT 3′) in PYP is not conserved among other PAS domain-containing proteins, interestingly, in three of the selected PAS domains containing proteins for *Legionella* genus, the sequences representing the PAS core are conserved (Figure S3). From the above observation, we could suggest that, although they are diverse in different organisms, PAS cores might be conserved in plants belonging to the same genus. We also performed a protein-protein interaction analysis, which revealed the possible interacting partners for PAS domain-containing proteins, which range from multisensor histidine kinase, CheA signal transduction histidine kinase, CheW protein, CheB methylesterase, 4-coumarate-CoA ligase, phenylalanine/histidine ammonia- lyase and Hpt sensor hybrid histidine kinase (Figure S7c and Table S1).

3.3.5. B_{12} Binding Domain

Many prokaryotes synthesize vitamin B_{12} (cobalamine) having a tetrapyrrole-like structure composed of a bound cobalt atom (Co) with two axial ligands. The lower ligand is known to be involved in vitamin B_{12} binding, while the upper one is used as a cofactor for different groups of enzymes/proteins, such as methyltransferases, reductases and isomerases [92]. The vitamin B_{12} binds to a specific domain, i.e., Asp/Glu-X-His-X-X-Gly-(41)-Ser/Thr-X-Leu-(26–28)-Gly-Gly, which is highly conserved in almost all B_{12}-dependent enzymes/proteins and characterized as a Rossmann fold, typically composed of 5 parallel β-sheets surrounded by 4–5 α helices [93,94]. Generally, vitamin B_{12} and its derivatives are known for their role in fatty acid and folate metabolism; however, recently they have been characterized as photoreceptors with a novel and unanticipated biological function as a light dependent transcriptional regulator [95]. Ortiz-Guerrero et al. [96] reported, for the first time, a light-induced excitation of CarH (Mer-like transcriptional factor/repressor) bound adenosylcobalamin ($AdoB_{12}$), which inhibited the formation of a stable CarH-AdenoB_{12} tetramer, thus allowing the gene expression for the carotenoid biosynthesis in *Myxobacteria*. In the dark, the stable CarH-AdoB$_{12}$ tetramer binds at the promoter region and shuts down the carotenoid biosynthesis [95–97]. Upon light illumination of the corrin ring associated with AdoB$_{12}$, it promotes the re-orientation of a helix bundle forming a covalent linkage between H132 and Co, and causes the CarH dissociation from the promoter region, which ultimately leads to the carotenoid gene expression [95,96,98]. Cheng et al. [92] also reported a small stand-alone B$_{12}$-binding domain protein, AerR in *Rhodobacter capsulatus*, which controls the light-dependent regulation of the biosynthesis of the photosystem via interacting with CrtJ, a repressor of the photosystem gene expression. Like CarH, the light illumination also leads to a covalent association between the His10 and Co ligand, which suggested that the light-dependent covalent linkage between the Co ligand and His residue might be the common mechanism in B$_{12}$-dependent photoreceptors. Furthermore, photoreceptors have also been naturally endowed with multiple photosensory domains, which could relay signals to output domains to control specific light-dependent functions [95]. In the present communication, we have selected a modular protein (accession number ABP71929.1) with the BLUF and vitamin B12 binding domain from *Rhodobacter sphaeroides* ATCC 17025 (Table 1).

We performed the alignment of the selected protein with the well-characterized B$_{12}$ binding domain containing proteins (CarH and AerR) and, surprisingly, observed that the crucial amino acids (Trp131, Val138, Glu141 and His142) involved in forming the binding pocket for AdoB$_{12}$ [95,99] were not observed in the selected protein sequence (Figure S4). Moreover, there are many prokaryotes which do not necessarily require the B$_{12}$ cofactor; hence, they can acquire alternative B$_{12}$-independent metabolic pathways for the same reaction [100]. The association of the B$_{12}$ domains with the BLUF domains was also reported by Cheng et al. [92], where the role of the B$_{12}$ domain is to sense light that is out of the absorption range of flavin. In several proteins, B12 binding domains are also found in combination with heme/oxygen sensing globin domains; however, none of these proteins have been characterized [92]. In proteins with combinations of histidine kinases or serine/threonine kinases with the B$_{12}$ domain, the kinases are responsible for regulating the response to light absorption by the B$_{12}$ domain [93].

The selected protein sequence (ABP71929.1) containing the B$_{12}$ domain was analyzed for a protein-protein interaction using String (version 11).The observation showed several interacting partners involved in the regulation of different signaling pathways (Figure S7d). The protein-protein interaction analysis revealed the interaction of the B$_{12}$ domain with the prephenate dehydratase enzyme (Rsph17025_0524; Figure S7d), which catalyzes the conversion of prephenate to phosphoenolpyruvate (PEP), water, carbon dioxide [101], and is generally involved in phenylalanine, tyrosine and tryptophan biosynthesis [102,103]. The selected protein also interacted with tyrosyl-tRNA synthetase (TyrS), an enzyme that catalyzes the attachment of tyrosine to tRNA in a two-step reaction. The interaction of the B$_{12}$ domain-containing protein with a transmembrane protein PA-phosphatase-like phosphodiesterase (Rsph17025_2732), a SARP family transcriptional regulator and a DNA mismatch repair protein, MutL, was also revealed through the protein-protein interaction analysis (Figure S7d). The BLUF-B$_{12}$ binding domain containing protein also interacted with the TonB protein, which communicates with outer membrane receptor proteins and drives the energy-dependent uptake of various substrates (such as iron citrate, enterochelin, aerobactin, etc.) into the periplasmic space [104].

3.3.6. PRK Superfamily

The PRK family represents a group of three types of P-loop containing kinases, i.e., phosphoribulokinase [105], uridine kinases [106], and pantothenate kinases (CoaA) [107]. This family is named after one of its members, i.e., phosphoribulokinase (PRK), which drives the phosphoryl transfer from Mg-ATP to ribulose 5-phosphate to form ribulose 1, 5-bisphosphate (RuBP) during the pentose phosphate pathway [105]. In *E. coli*, the *udk* gene encodes the pyrimidine salvage enzyme uridine kinase, causing the phosphorylation of uridine/cytidine into UMP/CMP using GTP as the phosphate donor [106]. Pantothenate kinase controls the rate-limiting step in the coenzyme A (CoA) biosynthesis pathway [107]. The *coaA* gene is transcribed to produce the 1.1 kb transcript, which is further translated into two protein products of 36.4 and 35.4 kDa, respectively [107]. The resultant proteins encoded by the *coaA* gene, showed a difference of eight amino acids at the N-terminus. The *E. coli* strains bearing multiple copies of the *coaA* gene showed a higher activity of the pantothenate kinase [107].

3.3.7. DNA pol 3 gamma3 Superfamily

In *E. coli*, DNA polymerase III (Pol III) is a complex holoenzyme comprised of three functionally distinct subassemblies, i.e., the core polymerase (α, ε, and θ subunit), the sliding clamp (β subunit) and the clamp loader complex (τ2γδδ′χψ subunit) [108]. The clamp loader is responsible for the DNA-dependent hydrolysis of ATP to load β2 clamps onto DNA for the interaction with core polymerases [109]. The gene *dnaX* encodes the ATP motor subunits of the clamp loader, i.e., one γ and two τ subunits, where the γ subunits are considered as a truncated product of the τ subunits [110,111]. The gamma (γ) subunit shares domains I-III with the tau (τ) subunit, while the domain IV and the entire alpha-interacting domain V subunit are only observed in the τ-subunit. The bacterial DNA pol III γ III domain and its homolog, the eukaryotic replication factor C (RFC), belong to the AAA-ATPase superfamily and are primarily involved in the breaking or restructuring of the supramolecular assembly of proteins [110]. In this communication, we have selected two BLUF modular domains in association with the DNA pol III γ III domain only (accession number AIQ92835.1) from *Methylobacterium oryzae* CBMB20, and the PRK11633-DNA pol III γ III domain (WP_048452447.1) from *Methylobacterium tarhaniae* (Table 1).

Furthermore, we have performed the alignment of the sequences representing the DNA pol III γ III domain against the well-characterized truncated sequence (1-373 amino acids) of the DNA polymerase III subunit gamma/tau (WP_113440333.1) from *E. coli* (Figure S5). The sequences corresponding to the DNA pol III γ III domains aligned against the domain II of the γ III subunit of DNA pol III from *E. coli*; however, most of the critical amino acids are not conserved amongst the DNA pol III γ III domain. In the *E. coli* DNA pol III γ III domain, Thr157, responsible for the hydrogen bond formation with the terminal phosphate of AMP-PNP, is only conserved in the DNA pol III γ III domain from *Methylobacterium tarhaniae* but not in *Methylobacterium oryzae* CBMB20 (Figure S5) [110]. Moreover, the C-terminal SARC motif (Ser168, Arg169, and Cys170) located in the α7 helix in the sensor1 region, which is highly conserved in the γ subunit of almost all organisms [112,113], is found missing in both of the DNA pol III γ III domains. Arg169 has dual roles, where on the one hand it acts as an "arginine finger" for the SARC motif, while on the other hand it is responsible for electrostatic and hydrophobic interactions which hold the δ subunit onto the γ III subunit. Arg215 is another critical amino acid and is a part of the conserved motif G/PxΦRXΦ (where Φ is any hydrophobic residue) located in the DNA pol III γ III domains, among prokaryotes as well as among eukaryotes [112]. The correct alignment of Arg215 is very crucial for its proper interaction with the phosphate group of ADP/ATP [110]. From the multiple sequence alignment analysis, it was observed that this particular amino acid is conserved in the BLUF-DNA pol III γ III domains but replaced by another polar amino acid (tyrosine:Q) in the BLUF- PRK11633-DNA pol III γ III domain (Figure S5).

The protein-protein interaction analysis revealed that most of the interacting partners of DNA pol III γ III (dnaX) are the components involved in the regulation of DNA replication, such as DNA pol I (Pol A), replicative DNA helicase (dnaB), DNA mismatch repair protein (recR), DNA pol III subunit α (dnaE), δ (holA and holB), ε (dnaQ), chi (holC), psi (holD) and β sliding clamp (dnaN) (Figure S7e and Table S1).

3.3.8. Cytochrome p450/ p450 Superfamily

Cytochrome p450s (CYPs) are a diverse group of heam-containing monooxygenases responsible for the oxidative degradation of steroids, fatty acid and xenobiotics [114]. These heam-thiolate proteins are named after their spectral absorbance peak at 450 nm, due to linkage with the cysteine thiolate of the protein [114]. In spite of a low sequence conservation, the structures are highly conserved. The cytochrome p450 core is made up of a four-helix bundle, helices J and K, two sets of beta-sheets, and it possesses a heam-binding loop, a proton-transfer groove and the conserved EXXR motif in helix K [115]. The hormone synthesis, cholesterol, and vitamin metabolism are some other pathways that are regulated by cytochrome p450s. In this study, we have selected a protein containing the BLUF domain associated with the p450 domain (accession number WP_045444510.1) from *Psychrobacter sp.* P11F6 (Table 1). A multiple sequence alignment of the sequence representing the p450 domain against the well-characterized CYP for *Bacillus subtilis* has been performed (Figure S6). The multiple sequence alignment revealed that, among two amino acids (i.e., arginine (Arg242) and proline (Pro243)), essential for substrate binding in the peroxygenase enzyme [116], arginine (Arg108) is also conserved in the BLUF regulated p450 domain (Figure S6). However, in the BLUF controlled p450 domain, hydrophobic proline is replaced by a polar amino acid residue, i.e., serine (Ser109). In several p450 enzymes, an adjacent acidic-polar amino acid pair was reported in the substrate binding site. In *Pseudomonas putida* camphor hydroxylase (CYP101A1), Asp251 and Thr252 were observed in the substrate binding site and used to relay protons onto iron-oxo species to activate the catalytic cycle [114]. The above observations indicate the different evolutionary routes adapted by these enzymes for the H_2O_2-driven catalysis. We also performed a protein-protein interaction analysis, which showed interacting partners for the selected protein (Figure S7f and Table S1). Most of the interacting proteins belonged to the fatty acid metabolism (CypB, CypC, CypD, and YitS) and to the secondary metabolism (PksJ and PksM).

3.3.9. REC Domain

Signal receiver (REC)/CheY-like photo-acceptor domains are the widely distributed regulatory domains in bacteria (CheY, OmpR, NtrC, and PhoB), however, they are now also reported in eukaryotes, for example, ETR1 from *Arabidopsis thaliana*. In the bacterial two-component regulatory system, the response regulator typically consists of a receiver domain that is covalently linked to an effector domain (DNA binding or catalytic units), and which is controlled by sensor kinase-catalyzed aspartyl phosphorylation [117]. The role of the REC domain is to receive the input signal perceived and transmitted from the sensor partner in the two-component systems. The REC domain interacts with different proteins to regulate processes like bacterial chemotaxis and some other regulatory pathways [118].

3.3.10. TetR and AcrR Domain

The TetR protein family is a group of transcriptional regulators with an HTH DNA-binding motif, which is widely distributed among bacteria [119]. TetR family proteins control efflux pumps and transporters having a role in antibiotic resistance and tolerance to toxic chemicals, synthesis of osmoprotectants, quorum sensing, drug resistance, virulence and sporulation [120,121]. However, the TetR family is named after its most characterized member, TetR, which has a role in the regulation of the expression of *tet* genes, involved in conferring tetracycline resistance in the bacterial system [120]. Proteins with the TetR domain maintain its optimal cellular level by feedback control. TetR (dimer) binds to two adjacent DNA major grooves (6bp each) located in the promoter region of the target gene on both of the strands, where helix $\alpha 3$ (Gln38 to His44) is involved in a sequence-specific recognition [119,120]. The Arg28 in helix $\alpha 2$ strengthens the specific contact with the complementary strand [119,120]. The hydrophobic core, developed from the contributing residues from the $\alpha 1$, $\alpha 2$, and $\alpha 3$ bundle, stabilizes the TetR DNA binding domain [119]. A highly conserved Lys48, located in $\alpha 4$,also has an essential role in the TetR-DNA complex formation [119].

The TetR family protein also works in a complex circuit with other proteins, including AcrR, a transcriptional repressor of the *acrAB* operon responsible for encoding a multidrug efflux pump which removes a wide range of antibiotics and confers antibiotic resistance in *E. coli* [121]. The AcrR protein (215 amino acid; dimer) crystal structure showed that it is composed of a three-helix DNA-binding domain and a unique C-terminal domain (large internal cavity) for ligand binding, which is structurally similar to members of the TetR family of transcriptional repressors [120,122]. It was predicted that ligand (rhodamine 6G, ethidium and proflavin) binding at the C-terminal ligand binding site leads to an alteration in the conformation of the N-terminal DNA binding region and thereby initiates transcription at the corresponding promoter of the target gene [123].

3.3.11. Endonuclease 3c and Endonuclease-NS Domain

The endonuclease 3c domain is widely distributed among the family of DNA repair proteins such as endonuclease III and DNA glycosylase (MutY or MBD4). The members of this family possess a conserved helix-hairpin-helix (HhH), a Gly/Pro-rich loop, as well as a conserved aspartate residue [124,125]. On the other hand, the endonuclease-NS domain has been explicitly reported in the DNA/RNA non-specific endonucleases and found both in prokaryotes and eukaryotes. The endonuclease-NS domain, containing endonucleases, showed an Mg^{2+} dependent cleavage of double-stranded as well as single-stranded nucleic acids. The extracellular *Serratia marcescens* nuclease is a well-characterized example of an endonuclease with an endonuclease-NS domain having a conserved histidine residue. The *Serratia marcescens* nuclease requires magnesium ion, three acidic (Asp107, Glu148 and Glu232) amino acid residues, as well as a few basic amino acid residues (Arg108, Arg152) for the endonuclease activity [126,127]. Proteins with the endonuclease-NS domain are broadly involved in hydrolase activity, nucleic acid binding and metal ion binding [126].

3.3.12. AraC Domain

The AraC protein (inducer/activator) regulates the *ara*BAD operon in *E. coli*, which is responsible for encoding structural components for the arabinose metabolism [128]. X-ray crystallization and NMR studies demonstrated that the AraC is a dimeric protein composed of two helix-turn-helix DNA-binding motifs [129]. AraC uses arabinose as a substrate, and the induction of the *ara*BAD operon depends on the concentration of extracellular arabinose ($>10^{-7}$ M) as well as on the rate of the arabinose uptake and catabolism [128].

3.3.13. Abhydrolase (α/β Hydrolase) Superfamily

The α/β hydrolase superfamily is a diverse group of hydrolytic enzymes that may differ in their catalytic function but that share a common fold with a conserved loop bearing catalytic triad [130]. The catalytic triad involves serine, glutamate/aspartate, and a histidine amino acid residue, and participates in the nucleophilic attack on a carbonyl carbon atom. The α/β hydrolase fold includes proteases, lipases, peroxidases, esterases, epoxide hydrolases and dehalogenases [131]. Unlike other proteins, the core of the protein belonging to this superfamily has an α/β sheet composed of 8 β strands lined to 6 α helices [130,132].

3.3.14. Domain of Unknown Function (DUF)

Generally, every protein domain has a distinct structure and function. However, there are several domains which have no known role, and these are referred to as domains of unknown function (DUFs). Most of the time these domains were ignored as having little relevance, but now there are reports which show that many DUFs are essential, as they are crucial for protein function. Basing themselves on sets of bioinformatic analyses of several uncharacterized DUFs, Goodcare et al. speculated about probable tasks which may be related to ATP binding or transcription [133].

3.3.15. ANK Repeats

Ankyrin (ANK) like repeats mediated protein-protein interactions between diverse groups of proteins [134,135] have been reported in almost all species [136]. The ANK proteins exhibit a domain shuffling via a horizontal gene transfer [137]. A protein may have several numbers of ANK repeats per protein [134,138]. Davis et al. [138] demonstrated the association of a specific 33 residue erythrocyte ankyrin repeat with an anion exchanger. A stack of ANK repeats has a superhelical arrangement with four consecutive repeats, and each unit contains two antiparallel helices and a beta-hairpin. ANK repeats may also occur in combinations with other types of domains [134].

3.3.16. RhoGEF Domain

The RhoGEF protein is a guanine nucleotide exchange factor (GEF) responsible for the activation of Rho family GTPases (Rho, Rac, and Cdc42) [139–141]; it controls a diverse array of cellular processes, including cellular differentiation [142], cell morphology [143], cell motility and adhesion [144], phagocytosis [145], cytokinesis [146], smooth muscle contraction [147], and the etiology of human disease such as hypertension [148] and cancer [149]. The Rho family proteins are generally found in two different conformational states, i.e., active GTP-bound and inactive GDP-bound [150]. The Rho family GTPases have a conserved domain of ~200 amino acid residues known as the RhoGEF domain or Dbl homology (DH) domain, which encodes a GEF specific to different Rho family members [151]. In addition to the RhoGEF domain, Rho family GTPases also have another functionally independent conserved domain (~100 amino acid residue), i.e., the pleckstrin homology (PH) domain, located at the C-terminus of the RhoGEF domain [152]. The C-terminal PH domain is generally involved in intracellular targeting and regulates the function of the RhoGEF domain. The RhoGEF domain has an α-helix bundle-like structure with three conserved regions, i.e., conserved region 1 (CR1), conserved region 2 (CR2) and conserved region 3 (CR3). Among these three conserved regions, CR1 and CR3 interact partly with α-6 and the DH/PH junction site, forming the Rho GTPase binding pocket.

3.3.17. PDZ Domain

PDZ domains or discs-large homologous regions (DHR) are widely spread in a wide range of membrane-bound signaling proteins from bacteria, yeasts, plants, insects and vertebrates [153,154]. The PDZ domain presents either as a single copy or as multiple copies and interacts either with the C-terminus of proteins or with internal peptide sequences [154]. Proteins with the PDZ domain are generally located at the plasma membrane, where they can directly interact with phosphatidylinositol 4, 5-bisphosphate (PIP2), as observed with the class II PDZ domain in syntenin [155]. PDZ domains (80–90 amino acids) are composed of compactly arranged six β-strands (βA-βF) and two α-helices (αA and αB) in a globular structure. PDZ domains interact with shaker-type K^+ channels in several MAGUKs or bind similar ligands of other transmembrane receptors [154].

3.3.18. GAL4-Fungal TF MHR Domain

Gal4 is a fungal-specific positive regulator for the expression of galactose-induced genes [156]. This domain is generally located at the N-terminus of several fungal-specific transcriptional regulators and contains a binuclear Zn cluster bound by six Cys residues; additionally, it is involved in the zinc-dependent binding of DNA [157,158]. The transcriptional regulators or proteins with the GAL4-fungal TF-MHR domain are generally involved in the arginine, proline, pyrimidine, quinate, maltose and galactose metabolisms, amide and GABA catabolism and leucine biosynthesis [159].

3.4. *BLUF Proteins for Optogenetic Tools*

Small photoreceptors, such as sensors of blue light using flavin (BLUF), light oxygen voltage (LOV) based receptors and cryptochromes, have been identified in the genomes of different organisms, from prokaryotes to higher eukaryotes, as has been shown in the present study and in several other studies [160,161]. As evident from the present study and other in vitro experiments [161–165], one of the peculiar features of these light-sensitive motifs are their association with various effector domains, thus hinting about the wide range of novel mechanistic and functional diversity controlled by light. This aspect is yet to be studied in detail. Investigations into PixJ1 (blue and green light), RcaE (red and green), changes in light-dependent *E. gracilis* PAC-α activity for *Drosophila*, behavioral modulation and neural responses in marine gastropod *Aplysia* and *Caenorhabditis elegans* [166–171] suggest sensitivities of each receptor toward a particular light bandwidth, which can be one of the important tools used to engineer a novel system for a broad range of physiological outputs. This study further expands the possibilities for the BLUF domain to be used as a powerful optogenetic tool for the development of novel optogenetic technologies. A vast variety of domain combinations of BLUF photoreceptors in different genomes (Figure 1) represents a promising and valuable tool to design novel photo-regulated enzymes, messengers, photo-modulation of gene expression patterns, photo-control of the virulence in pathogenic bacteria through the recombinant expression of such systems, photobehavioural responses in photobacteria, modulation of neural systems and dynamic molecular switches to regulate biological activities. BLUF domains, in combination with the EAL, GGDEF or CHD domains, could be utilized for the photo-dependent regulation of c-di-GMP and cAMP associated signaling in bacteria [64,85]. The BLUF domain associated with a B_{12} binding domain has also been analyzed; however, it does not show the important amino acids that are required in order to form the binding pocket for $AdoB_{12}$. Cheng et al. [92] also reported a similar BLUF module and suggested that the associated B_{12} binding domain also has a photosensory function which can regulate activity in response to light. The B_{12} binding domain broadens the absorption range for the BLUF photosensor, which could be critical to several regulatory pathways [96,172]. We could also use this modular combination for the photo-dependent regulation of pathways like carotenoid synthesis or photosystem biosynthesis [92]. Another combination which could be engineered and used for an optogenetic application is the association of the BLUF domain with the DNA pol III γ III domain. Using this modular architecture, we could regulate the actions of different components involved in DNA replication in a light-dependent manner. Furthermore, the BLUF domain is also found in association with the p450 (cytochrome p450) domain, which could be used as an optogenetic tool for the light-dependent regulation of several pathways like fatty acid metabolism and secondary metabolism. The optogenetic potential of the BLUF domain could also be acquired in the two-component regulatory system for both prokaryotes and eukaryotes. A modular architecture in which the BLUF sensor domain is associated with the Rec (receiver) domain could be exploited as a two-component regulatory system for the photo-dependent regulation of processes like bacterial chemotaxis and other regulatory pathways [118]. The optogenetic potential of the BLUF domain could also be extended to the regulation of the efflux pump and transporter involved in antibiotic resistance, tolerance to a toxic chemical, synthesis of an osmoprotectant, quorum sensing, drug resistance and sporulation [120–122]. The BLUF domain associated with the TetR or AcrR domain may be exploited for the light-dependent regulation of pathways (as mentioned earlier) in bacteria. The BLUF domain was also analyzed with the endonuclease 3c and endonuclease_NS domains for their optogenetic potential. The BLUF domain endonuclease 3c could be engineered and used as an optogenetic tool for the light-dependent regulation of the DNA repair process. On the other hand, the BLUF domain associated with the endonuclease_NS domain could be exploited for the light-dependent regulation of processes like hydrolase activity, nucleic acid binding and metal ion binding [126]. The modular architectures comprised of BLUF associated with the AraC and abhydrolase domains could also be harnessed for the light-dependent regulation of the arabinose metabolism, as well as the diverse group of hydrolytic enzymes that include proteases, lipases, peroxidases, esterases, epoxide hydrolases, and dehalogenases, respectively [131]. BLUF in combination with the RhoGEF domain is also considered an important modular architecture, which could be used as an optogenetic tool for the light-dependent regulation of a diverse array of cellular processes [142–146,148,149].

4. Conclusions

Applications of the photoreceptors in order to quickly control molecular machines and, in turn, biological systems and processes, present the scientific community with an exciting opportunity with several possibilities, approaches, along with their limitations as well. Recently, several such approaches have been reported, including photoswitches, UV photo-reactivation and deactivation, spectral tuning, and photocaging [25]. However, a detailed understanding of electron transfer mechanisms, transient state intermediates, and amino acid patterns will allow the development of more precise recombinant techniques, fusion proteins and complexes for improving such systems and providing an alternate route to design broadly reactive light-sensitive probes. In the case of LOV domains, the deprotonation of flavin N (5) involves rate-determination for the recovery, using base catalysis, pH, proton inventory and structural studies [173]. It will also be interesting to study mechanisms to convert these small spectral shifts into more significant jumps leading to many fold increases in the rage of 400–500 nm in the activities of such receptors and downstream signals. Theoretically, the fold increase in the signal response could be improved to that extent [173,174]. Biocatalytic reactions using photoactivated enzymes, produced either through recombinant methods or through directed evolution, can reduce the complexity of the system by controlling the system remotely to deliver high-value materials and compounds in biotechnology or the pharma industry. Photo-controlled receptors like BLUF can play a vital role in biotransformation cascade in the same way as that of photosensitive chemical groups like O-nitrobenzyl, 3-nitrophenyl and benzyloxycarbonylphenylin organic reaction steps, where the reaction's product acts as the substrate for the next reaction in a multistep pathway [175]. We have identified and analysed 34 such proteins containing the BLUF domain in association with different effector domains, such as kinases, phosphatases, phosphodiesterases, a transcriptional repressor (TetR), PAS, endonuclease, PBP proteins, etc., involved in regulating a wide range of cellular processes. All of the selected proteins have a conserved catalytic core, including tyrosine, glutamine, and tryptophan or methionine, which are essential for the BLUF photo-activation and photocycle. Until now, several photoreceptors, such as channelrhodopsin2 (ChR2) [176], phytochrome [177], cryptochromes [178], LOV [19], as well as BLUF [4,30], have been adopted for their optogenetic potentials. However, considering the modularities in the BLUF domain architecture and their unexplored nature, these combinations have a great potential to befurther utilized for the development of novel optogenetic technologies. To conclude, these photoactivated/controlled systems could be the way forward in synthetic biology, as different and subtle differences in the light sensitivities of these vast arrays of receptors can be harnessed to regulate a reaction cascade in an engineered organism by choosing a particular photoreactivity and control.

Supplementary Materials: The following are available online at http://www.mdpi.com/2076-3417/9/18/3924/s1, Figure S1: Multiple sequence alignment of the BLUF coupled EAL domain. Sequence representing the EAL domain from the BlrP1 protein was used as the template for the sequence alignment analysis. Amino acid residues in solid box are the conserved residues involved in the formation of the EAL active site, Figure S2: Multiple sequence alignment of the BLUF coupled CHD domain. Sequence representing the CHD domain from the bPAC protein was used as template for the sequence alignment analysis. Amino acid residues in solid box are the conserved residues involved in the formation of the nucleotide binding site, Figure S3: Multiple sequence alignment of the BLUF coupled PAS domain. Sequence representing the PAS domain from the photoactivated yellow protein (PYP) from *Halorhodospira halophila* was used as template for the sequence alignment analysis. Amino acid residues in solid box are representing the PAS core motif responsible for the generation and propagation of the signal to the adjoining effector domain, Figure S4: Multiple sequence alignment of the BLUF coupled vitamin B_{12} binding domain. Sequence representing the B_{12} binding domain from the CarH and AerR proteins was used as template for the sequence alignment analysis. The conserved amino acids (Trp 131, Val138, Glu141 and His142) essential for forming the binding pocket for the substrate, i.e., $AdoB_{12}$, is not found in the aligned portion of the BLUF coupled vitamin B_{12} binding domain, Figure S5: Multiple sequence alignment of the BLUF coupled DNA pol III γ III domain. The sequence of the well characterized truncated (1-373 amino acids) DNA polymerase III subunit gamma/tau (WP_113440333.1) from *E. coli* was used as template for the sequence alignment analysis. Amino acid residues in solid box represent the important residues crucial for the enzyme activity, Figure S6: Multiple sequence alignment of the BLUF coupled p450 and the well characterized CYP protein from *Bacillus subtilis* (used as template). Amino acid residues in solid box representing the conserved (Arg) and altered (Pro to Ser) amino acid residues essential for the substrate binding, Figure S7: Protein-Protein interaction network depicting interacting partners of the selected effector domain (EAL, CHD, PAS, B_{12}, DNA POL III γ III and p450) of the BLUF modular proteins. Protein highlighted in yellow is the query protein. Protein-protein interaction analysis was performed by using String version 11 (https://string-db.org/). Details of the query proteins, domains, along with the annotations are given in Table S1, Table S1: Output showing the details of query proteins, domains, interacting proteins and annotated functions.

Author Contributions: M.S.K. had analysed all the sequences in details and wrote the manuscript. R.S. had done preliminary analysis of the BLUF coupled proteins. S.K.V. had conceptualized protein-protein interaction of the BLUF coupled domains and wrote manuscript. S.K.S. had done detailed structure-function analysis of the BLUF and its modular domains and wrote relevant part of the paper. S.K. had conceptualized, outlined and wrote the manuscript.

Funding: M.S.K. research associateship was supported from SR/NM/NB-1087/2017(G)-JNU. S.K. was supported by the SR/NM/NB-1087/2017(G)-JNU. S.K.S. was supported by the research endowment fund from MUJ.

Conflicts of Interest: The authors have no conflict of interest.

References

1. Purcell, E.B.; Crosson, S. Photoregulation in prokaryotes. *Curr. Opin. Microbiol.* **2008**, *11*, 168–178. [CrossRef] [PubMed]
2. Ohlendorf, R.; Vidavski, R.R.; Eldar, A.; Moffat, K.; Moglich, A. From dusk till dawn: One-plasmid systems for light-regulated gene expression. *J. Mol. Biol.* **2012**, *416*, 534–542. [CrossRef] [PubMed]
3. Stierl, M.; Stumpf, P.; Udwari, D.; Gueta, R.; Hagedorn, R.; Losi, A.; Gärtner, W.; Petereit, L.; Efetova, M.; Schwarzel, M.; et al. Light modulation of cellular cAMP by a small bacterial photoactivated adenylyl cyclase, bPAC, of the soil bacterium *Beggiatoa*. *J. Biol. Chem.* **2011**, *286*, 1181–1188. [CrossRef] [PubMed]
4. Barends, T.R.M.; Hartmann, E.; Griese, J.J.; Beitlich, T.; Kirienko, N.V.; Ryjenkov, D.A.; Reinstein, J.; Shoeman, R.L.; Gomelsky, R.; Schlichting, I. Structure and mechanism of a bacterial light-regulated cyclic nucleotide phosphodiesterase. *Nature* **2009**, *459*, 1015–1020. [CrossRef] [PubMed]
5. Masuda, S.; Bauer, C.E. AppA is a blue light photoreceptor that antirepresses photosynthesis gene expression in *Rhodobacter sphaeroides*. *Cell* **2002**, *110*, 613–623. [CrossRef]
6. Tanwar, M.; Nahar, S.; Gulati, S.; Veetil, S.K.; Kateriya, S. Molecular determinant modulates thermal recovery kinetics and structural integrity of the bacterial BLUF photoreceptor. *FEBS Lett.* **2016**, *590*, 2146–2157. [CrossRef] [PubMed]
7. Tanaka, K.; Nakasone, Y.; Okajima, K.; Ikeuchi, M.; Tokutomi, S.; Terazima, M. Oligomeric-state dependent conformational change of the BLUF protein TePixD (Tll0078). *J. Mol. Biol.* **2009**, *386*, 1290–1300. [CrossRef]
8. Tanaka, K.; Nakasone, Y.; Okajima, K.; Ikeuchi, M.; Tokutomi, S.; Terazima, M. Light-induced conformational change and transient dissociation reaction of the BLUF photoreceptor *Synechocystis* PixD (Slr1694). *J. Mol. Biol.* **2011**, *409*, 773–785. [CrossRef]
9. Yuan, H.; Dragnea, V.; Wu, Q.; Gardner, K.H.; Bauer, C.E. Mutational and structural studies of the PixD BLUF output signal that affects light-regulated interactions with PixE. *Biochemistry* **2011**, *50*, 6365–6375. [CrossRef]
10. Laan, W.; Bednarz, T.; Heberle, J.; Hellingwerf, K.J. Chromophore composition of a heterologously expressed BLUF-domain. *Photochem. Photobiol. Sci.* **2004**, *3*, 1011–1016. [CrossRef]
11. Kennis, J.T.M.; Mathes, T. Molecular eyes: Proteins that transform light into biological information. *Interface Focus* **2013**, *3*, 20130005. [CrossRef] [PubMed]
12. Laan, W.; van der Horst, M.A.; van Stokkum, I.H.; Hellingwerf, K.J. Initial characterization of the primary photochemistry of AppA, a blue-light-using flavin adenine dinucleotide-domain containing transcriptional antirepressor protein from *Rhodobacter sphaeroides*: A key role for reversible intramolecular proton transfer from the flavin adenine dinucleotide chromophore to a conserved tyrosine? *Photochem. Photobiol.* **2003**, *78*, 290–297. [PubMed]
13. Kraft, B.J.; Masuda, S.; Kikuchi, J.; Dragnea, V.; Tollin, G.; Zaleski, J.M.; Bauer, C.E. Spectroscopic and mutational analysis of the blue-light photoreceptor AppA: A novel photocycle involving flavin stacking with an aromatic amino acid? *Biochemistry* **2003**, *42*, 6726–6734. [CrossRef] [PubMed]
14. Unno, M.; Masuda, S.; Ono, T.-A.; Yamauchi, S. Orientation of a key glutamine residue in the BLUF domain from AppA revealed by mutagenesis, spectroscopy, and quantum chemical calculations. *J. Am. Chem. Soc.* **2006**, *128*, 5638–5639. [CrossRef]
15. Penzkofer, A.; Tanwar, M.; Veetil, S.K.; Kateriya, S.; Stierl, M.; Hegemann, P. Photo-dynamics and thermal behavior of the BLUF domain containing adenylate cyclase NgPAC2 from the amoeboflagellate *Naegleria gruberi* NEG-M strain. *Chem. Phys.* **2013**, *412*, 96–108. [CrossRef]
16. Bonetti, C.; Mathes, T.; van Stokkum, I.H.M.; Mullen, K.M.; Groot, M.-L.; van Grondelle, R.; Hegemann, P.; Kennis, J.T.M. Hydrogen-bond switching among flavin and amino acid side chains in the BLUF photoreceptor observed by ultrafast infrared spectroscopy. *Biophys. J.* **2008**, *95*, 4790–4802. [CrossRef] [PubMed]
17. Gauden, M.; van Stokkum, I.H.M.; Key, J.M.; Luhrs, D.C.; van Grondelle, R.; Hegemann, P.; Kennis, J.T.M. Hydrogen-bond switching through a radical pair mechanism in a flavin-binding photoreceptor. *Proc. Natl. Acad. Sci. USA* **2006**, *103*, 10895–10900. [CrossRef] [PubMed]

18. Han, Y.; Braatsch, S.; Osterloh, L.; Klug, G. A eukaryotic BLUF domain mediates light-dependent gene expression in the purple bacterium *Rhodobacter sphaeroides* 2. 4. 1. *Proc. Natl. Acad. Sci. USA* **2004**, *101*, 12306–12311. [CrossRef]
19. Wang, X.; Chen, X.J.; Yang, Y. Spatiotemporal control of gene expression by a light-switchable transgene system. *Nat. Methods* **2012**, *9*, 266–269. [CrossRef]
20. Strickland, D.; Moffat, K.; Sosnick, T.R. Light activated DNA binding in a designed allosteric protein. *Proc. Natl. Acad. Sci. USA* **2008**, *105*, 10709–10714. [CrossRef]
21. Moglich, A.; Moffat, K. Engineered photoreceptors as novel optogenetic tools. *Photochem. Photobiol. Sci.* **2010**, *9*, 1286–1300. [CrossRef] [PubMed]
22. Moglich, A.; Ayers, R.A.; Moffat, K. Design and signaling mechanism of light-regulated histidine kinases. *J. Mol. Biol.* **2009**, *385*, 1433–1444. [CrossRef] [PubMed]
23. Wu, Y.I.; Frey, D.; Lungu, O.I.; Jaehrig, A.; Schlichting, I.; Kuhlman, B.; Hahn, K.M. A genetically encoded photoactivatable Rac controls the motility of living cells. *Nature* **2009**, *461*, 104–108. [CrossRef] [PubMed]
24. Polstein, L.R.; Gersbach, C.A. Light-inducible spatiotemporal gene regulation using engineered transcription factors. *Mol. Ther.* **2012**, *20*, 193–194.
25. Krauss, U.; Drepper, T.; Jaeger, K.E. Enlightened enzymes: Strategies to create novel photoresponsive proteins. *Chem. Eur. J.* **2011**, *17*, 2552–2560. [CrossRef] [PubMed]
26. Hisert, K.B.; MacCoss, M.; Shiloh, M.U.; Darwin, K.H.; Singh, S.; Jones, R.A.; Ehrt, S.; Zhang, Z.; Gaffney, B.L.; Gandotra, S.; et al. A glutamate-alanine-leucine (EAL) domain protein of *Salmonella* controls bacterial survival in mice, antioxidant defence and killing of macrophages: Role of cyclic diGMP. *Mol. Microbiol.* **2005**, *56*, 1234–1245. [CrossRef] [PubMed]
27. Hoffman, L.R.; D'argenio, D.A.; MacCoss, M.J.; Zhang, Z.; Jones, R.A.; Miller, S.I. Aminoglycoside antibiotics induce bacterial biofilm formation. *Nature* **2005**, *436*, 1171–1175. [CrossRef]
28. Kulesekara, H.; Lee, V.; Brencic, A.; Liberati, N.; Urbach, J.; Miyata, S.; Lee, D.G.; Neely, A.N.; Hyodo, M.; Hayakawa, Y.; et al. Analysis of *Pseudomonas aeruginosa* diguanylate cyclases and phosphodiesterases reveals a role for bis (39-59)-cyclic-GMP in virulence. *Proc. Natl. Acad. Sci. USA* **2006**, *103*, 2839–2844. [CrossRef]
29. Claret, L.; Miquel, S.; Vieille, N.; Ryjenkov, D.A.; Gomelsky, M.; Darfeuille-Michaud, A. The flagellar sigma factor FliA regulates adhesion and invasion of Crohn disease-associated *Escherichia coli* via a cyclic dimeric GMP-dependent pathway. *J. Biol. Chem.* **2007**, *282*, 33275–33283. [CrossRef]
30. Tschowri, N.; Busse, S.; Hengge, R. The BLUF-EAL protein YcgF acts as a direct anti-repressor in a blue-light response of *Escherichia coli*. *Gene Dev.* **2009**, *23*, 522–534. [CrossRef]
31. Kader, A.; Simm, R.; Gerstel, U.; Morr, M.; Römling, U. Hierarchical involvement of various GGDEF domain proteins in rdar morphotype development of *Salmonella enteric* serovar Typhimurium. *Mol. Microbiol.* **2006**, *60*, 602–616. [CrossRef]
32. Wei, C.; Jiang, W.; Zhao, M.; Ling, J.; Zeng, X.; Deng, J.; Jin, D.; Dow, J.M.; Sun, W. A systematic analysis of the role of GGDEF-EAL domain proteins in virulence and motility in *Xanthomonas oryzae* pv. *Oryzicola*. *Sci. Rep.* **2016**, *6*, 23769. [CrossRef]
33. Lindner, R.; Hartmann, E.; Tarnawski, M.; Winkler, A.; Frey, D.; Reinstein, J.; Meinhart, A.; Schlichting, I. Photoactivation Mechanism of a Bacterial Light-Regulated Adenylyl Cyclase. *J. Mol. Biol.* **2017**, *429*, 1336–1351. [CrossRef] [PubMed]
34. Penzkofer, A.; Tanwar, M.; Veetil, S.K.; Kateriya, S. Photo-dynamics of photoactivated cyclase LiPAC from the spirochetebacterium *Leptonema illini* strain 3055^{T}. *Trends Appl. Spectrol.* **2014**, *11*, 40–62.
35. Iseki, M.; Matsunaga, S.; Murakami, A.; Ohno, K.; Shiga, K.; Yoshida, K.; Sugai, M.; Takahashi, T.; Hori, T.; Watanabe, M. A blue-light-activated adenylyl cyclase mediates photoavoidance in *Euglena gracilis*. *Nature* **2002**, *415*, 1047–1051. [CrossRef] [PubMed]
36. Fritz-Laylin, L.K.; Prochnik, S.E.; Ginger, M.L.; Dacks, J.B.; Carpenter, M.L.; Field, M.C.; Kuo, A.; Paredez, A.; Chapman, J.; Pham, J.; et al. The genome of *Naegleria gruberi* illuminates early eukaryotic versatility. *Cell* **2010**, *140*, 631–642. [CrossRef] [PubMed]
37. Ryu, M.-H.; Moskvin, O.V.; Siltberg-Liberles, J.; Gomelsky, M. Natural and engineered photoactivated nucleotidylcyclases for optogenetic applications. *J. Biol. Chem.* **2010**, *285*, 41501–41508. [CrossRef] [PubMed]
38. Ohki, M.; Sugiyama, K.; Kawai, F.; Tanaka, H.; Nihei, Y.; Unzai, S.; Takebe, M.; Matsunaga, S.; Adachi, S.I.; Shibayama, N.; et al. Structural insight into photoactivation of an Adenylate cyclase from a photosynthetic cyanobacterium. *Proc. Natl. Acad. Sci. USA* **2016**, *113*, 6659–6664. [CrossRef] [PubMed]

39. Taylor, B.L.; Zhulin, I.B. PAS domains: Internal sensors of oxygen, redox potential, and light. *Microbiol. Mol. Biol. Rev.* **1999**, *63*, 479–506. [PubMed]
40. Baca, M.; Borgstahl, G.E.; Boissinot, M.; Burke, P.M.; Williams, D.R.; Slater, K.A.; Getzoff, E.D. Complete chemical structure of photoactive yellow protein: Novel thioester-linked 4-hydroxycinnamyl chromophore and photocycle chemistry. *Biochemistry* **1994**, *33*, 14369–14377. [CrossRef] [PubMed]
41. Giraud, E.; Fardoux, J.; Fourrier, N.; Hannibal, L.; Genty, B.; Bouyer, P.; Dreyfus, B.; Verméglio, A. Bacteriophytochrome controls photosystem synthesis in anoxygenic bacteria. *Nature* **2002**, *417*, 202. [CrossRef] [PubMed]
42. Montgomery, B.L. Sensing the light: Photoreceptive systems and signal transduction in cyanobacteria. *Mol. Microbiol.* **2007**, *64*, 16–27. [CrossRef] [PubMed]
43. Penzkofer, A.; Kateriya, S.; Hegemann, P. Photodynamics of the optogenetic BLUF coupled photoactivated adenylyl cyclases (PACs). *Dyes Pigments* **2016**, *135*, 102–112. [CrossRef]
44. Christie, J.M.; Gawthorne, J.; Young, G.; Fraser, N.J.; Roe, A.J. LOV to BLUF: Flavoprotein Contributions to the Optogenetic Toolkit. *Mol. Plant* **2012**, *5*, 533–544. [CrossRef] [PubMed]
45. Losi, A.; Gardner, K.H.; Moglich, A. Blue-Light Receptors for optogentetics. *Chem. Rev.* **2018**, *118*, 10659–10709. [CrossRef] [PubMed]
46. Geer, L.Y.; Domrachev, M.; Lipman, D.J.; Bryant, S.H. CDART: Protein homology by domain architecture. *Genome Res.* **2002**, *12*, 1619–1623. [CrossRef]
47. Hall, A. BioEdit: A user-friendly biological sequence alignment editor and analysis program for window 95/98/NT. *Nucliec Acid Res.* **1999**, *41*, 95–98.
48. Bailey, T.L.; Boden, M.; Buske, F.A.; Frith, M.; Grant, C.E.; Clementi, L.; Ren, J.; Li, W.W.; Nobel, W.S. MEME SUITE: Tools for motif discovery and searching. *Nucleic Acid Res.* **2009**, *37*, 202–208. [CrossRef]
49. Szklarczyk, D.; Morris, J.H.; Cook, H.; Kuhn, M.; Wyder, S.; Simonovic, M.; Santos, A.; Doncheva, N.T.; Roth, A.; Bork, P.; et al. The STRING database in 2017: Quality-controlled protein-protein association networks, made broadly accessible. *Nucleic Acids Res.* **2016**, *45*, 362–368. [CrossRef]
50. Jones, D.T.; Taylor, W.R.; Thornton, J.M. The rapid generation of mutation data matrices from protein sequences. *Comp. Appl. Biosci.* **1992**, *8*, 275–282. [CrossRef]
51. Tamura, K.; Stecher, G.; Peterson, D.; Filipski, A.; Kumar, S. MEGA6: Molecular evolutionary genetics analysis version 6. 0. *Mol. Biol. Evol.* **2013**, *30*, 2725–2729. [CrossRef] [PubMed]
52. Omasits, U.; Ahrens, C.H.; Müller, S.; Wollscheid, B. Protter: Interactive protein feature visualization and integration with experimental proteomic data. *Bioinformatics* **2014**, *30*, 884–886. [CrossRef] [PubMed]
53. Kelley, L.A.; Mezulis, S.; Yates, C.M.; Wass, M.N.; Sternberg, M.J.E. The Phyre2 web portal for protein modeling, prediction and analysis. *Nat. Protoc.* **2015**, *10*, 845–858. [CrossRef] [PubMed]
54. Roy, A.; Kucukural, A.; Zhang, Y. I-TASSER: A unified platform for automated protein structure and function prediction. *Nat. Protoc.* **2010**, *5*, 725–738. [CrossRef] [PubMed]
55. Hooft, R.W.; Vriend, G.; Sander, C.; Abola, E.E. Errors I protein structures. *Nature* **1996**, *381*, 272. [CrossRef] [PubMed]
56. Laskowski, R.A.; MacArthur, M.W.; Moss, D.S.; Thornton, J.M. PROCHECK: A program to check the stereochemical quality of protein structures. *J. Appl. Crystallogr.* **1993**, *26*, 283–291. [CrossRef]
57. Gauden, M.; Yeremenko, S.; Laan, W.; van Stokkum, I.H.M.; Ihalainen, J.A.; van Grondelle, R.; Hellingwerf, K.J.; Kennis, J.T.M. Photocycle of the flavin-binding photoreceptor appa, a bacterial transcriptional antirepressor of photosynthesis genes. *Biochemistry* **2005**, *44*, 3653–3662. [CrossRef]
58. Fukushima, Y.; Okajima, K.; Shibata, Y.; Ikeuchi, M.; Itoh, S. Primary intermediate in the photocycle of a blue-light sensory BLUF FAD-protein, Tll0078, of *Synechococcus elongatus* BP-1. *Biochemistry* **2005**, *44*, 5149–5158. [CrossRef]
59. Moglich, A.; Ayers, R.A.; Moffat, K. Addition at the molecular level: Signal integration in designed Per-ARNT-Sim receptor proteins. *J. Mol. Biol.* **2010**, *400*, 477–486. [CrossRef]
60. Moglich, A.; Yang, X.; Ayers, R.A.; Moffat, K. Structure and function of plant photoreceptors. *Annu. Rev. Plant Biol.* **2010**, *61*, 21–47. [CrossRef]
61. Christie, J.M.; Swartz, T.E.; Bogomolni, R.A.; Briggs, W.R. Phototropin LOV domains exhibit distinct roles in regulating photoreceptor function. *Plant J.* **2002**, *32*, 205–219. [CrossRef] [PubMed]
62. Christie, J.M. Phototropin blue-light receptors. *Ann. Rev. Plant Biol.* **2007**, *58*, 21–45. [CrossRef] [PubMed]
63. Delano, W.L. *PyMOL Molecular Viewer*; Delano Scientific: Palo Alto, CA, USA, 2008.

64. Galperin, M.Y.; Nikolskaya, A.N.; Koonin, E.V. Novel domains of the prokaryotic two-component signal transduction systems. *FEMS Microbiol. Lett.* **2001**, *203*, 11–21. [CrossRef] [PubMed]
65. Mills, E.; Pultz, I.S.; Kulasekara, H.D.; Miller, S.I. The bacterial second messenger c-di-GMP: Mechanisms of signaling. *Cell. Microbiol.* **2011**, *13*, 1122–1129. [CrossRef] [PubMed]
66. Sondermann, H.; Shikuma, N.J.; Yildiz, F.H. You've come a long way: C-di-GMP signaling. *Curr. Opin. Microbiol.* **2012**, *13*, 140–146. [CrossRef] [PubMed]
67. Malone, J.; Williams, R.; Christen, M.; Jenal, U.; Spiers, A.J.; Rainey, P.B. The structure–function relationship of WspR, a *Pseudomonas fluorescens* response regulator with a GGDEF output domain. *Microbiology* **2007**, *153*, 980–994. [CrossRef]
68. Wassmann, P.; Chan, C.; Paul, R.; Beck, A.; Heerklotz, H.; Jenal, U.; Schirmer, T. Structure of BeF3-modified response regulator PleD: Implications of diguanylate cyclase activation, catalysis, and feedback inhibition. *Structure* **2007**, *15*, 915–927. [CrossRef] [PubMed]
69. Rao, F.; Yang, Y.; Qi, Y.; Liang, Z.X. Catalytic mechanism of c-di-GMP specific phosphodiesterase: A study of the EAL domain-containing RocR from *Pseudomonas aeruginosa*. *J. Bacteriol.* **2008**, *190*, 3622–3631. [CrossRef]
70. Qi, Y.; Rao, F.; Luo, Z.; Liang, Z.X. A flavin cofactor-binding PAS domain regulates c-di-GMP synthesis in AxDGC2 from *Acetobacter xylinum*. *Biochemistry* **2009**, *48*, 10275–10285. [CrossRef]
71. Christen, M.; Christen, B.; Folcher, M.; Schauerte, A.; Jenal, U. Identification and characterization of a cyclic di-GMP-specific phosphodiesterase and its allosteric control by GTP. *J. Biol. Chem.* **2005**, *285*, 30829–30837. [CrossRef]
72. Newell, P.D.; Yoshioka, S.; Hvorecny, K.L.; Monds, R.D.; O'Toole, G.A. Systematic analysis of diguanylate cyclases that promote biofilm formation by *Pseudomonas fluorescens* Pf0-1. *J. Bacteriol.* **2011**, *193*, 4685–4698. [CrossRef] [PubMed]
73. Newell, P.D.; Monds, R.D.; O'Toole, G.A. LapD is a bis-(3′,5′)-cyclic dimeric GMP-binding protein that regulates surface attachment by *Pseudomonas fluorescens* Pf0-1. *Proc. Natl. Acad. Sci. USA* **2009**, *106*, 3461–3466. [CrossRef] [PubMed]
74. Hengge, R.; Galperin, M.Y.; Ghigo, J.M.; Gomelsky, M.; Green, J.; Hughes, K.T.; Jenal, U.; Landini, P. Systematic nomenclature for GGDEF and EAL domain-containing cyclic di-GMP turnover proteins of Escherichia coli. *J. Bacteriol.* **2016**, *198*, 7–11. [CrossRef] [PubMed]
75. Yang, F.; Tian, T.; Li, X.; Fan, S.; Chen, H.; Wu, M.; Yang, C.-H.; He, C. The degenerate EAL GGDEF Domain protein filp functions as a cyclic di-gmp receptor and specifically interacts with the PilZ-domain protein PXO_02715 to regulate virulence in *Xanthomonas oryzae* pv. *oryzae*. *MPMI* **2014**, *27*, 578–589. [CrossRef] [PubMed]
76. Suzuki, K.; Babitzke, P.; Kushner, S.R.; Romeo, T. Identification of a novel regulatory protein (CsrD) that targets the global regulatory RNAs CsrB and CsrC for degradation by RNase E. *Genes Dev.* **2006**, *20*, 2605–2617. [CrossRef] [PubMed]
77. Wu, Q.; Gardner, K.H. Structure and Insight into Blue Light-Induced Changes in the BlrP1 BLUF Domain. *Biochemistry* **2009**, *48*, 2620–2629. [CrossRef]
78. Yuan, H.; Anderson, S.; Masuda, S.; Dragnea, V.; Moffat, K.; Bauer, C. Crystal structures of the Synechocystis photoreceptor Slr1694 reveal distinct structural states related to signaling. *Biochemistry* **2006**, *45*, 12687–12694. [CrossRef]
79. Jung, A.; Domratcheva, T.; Tarutina, M.; Wu, Q.; Ko, W.H.; Shoeman, R.L.; Gomelsky, M.; Gardner, K.H.; Schlichting, I. Structure of a bacterial BLUF photoreceptor: Insights into blue light-mediated signal transduction. *Proc. Natl. Acad. Sci. USA* **2005**, *102*, 12350–12355. [CrossRef]
80. Kita, A.; Okajima, K.; Morimoto, Y.; Ikeuchi, M.; Miki, K. Structure of a cyanobacterial BLUF protein, Tll0078, containing a novel FAD-binding blue light sensor domain. *J. Mol. Biol.* **2005**, *349*, 1–9. [CrossRef]
81. Wood, T.K. Insights on *Escherichia coli* biofilm formation and inhibition from whole-transcriptome profiling. *Environ. Microbiol.* **2009**, *11*, 1–15. [CrossRef]
82. Zucca, S.; Pasotti, L.; Politi, N.; Casanova, M.; Mazzini, G.; Cusella De Angelis, M.G.; Magni, P. Multi-faceted characterization of a novel LuxR repressible promoter library for *Escherichia coli*. *PLoS ONE* **2015**, *10*, e0126264. [CrossRef] [PubMed]
83. Kendall, M.M.; Sperandio, V. Cell-to-cell signaling in *E. coli* and *Salmonella*. *Eco Sal Plus* **2014**, *6*, 1.

84. Kim, S.K.; Kimura, S.; Shinagawa, H.; Nakata, A.; Lee, K.S.; Wanner, B.L.; Makino, K. Dual transcriptional regulation of the *Escherichia coli* phosphate-starvation-inducible *psiE* gene of the phosphate regulon by phob and the cyclic AMP (cAMP)-cAMP receptor protein complex. *J. Bacteriol.* **2000**, *182*, 5596–5599. [CrossRef] [PubMed]
85. Linder, J.U.; Schultz, J.E. The class III adenylyl cyclases: Multi-purpose signaling modules. *Cell. Signal.* **2003**, *15*, 1081–1089. [CrossRef]
86. Sinha, S.C.; Sprang, S.R. Structures, mechanism, regulation and evolution of class III nucleotidylcyclases. In *Reviews of Physiology Biochemistry and Pharmacology*; Springer: Berlin/Heidelberg, Germany, 2006; pp. 105–140.
87. Vercellino, I.; Rezabkova, L.; Olieric, V.; Polyhach, Y.; Weinert, T.; Kammerer, R.A.; Jeschke, G.; Korkhova, V.M. Role of the nucleotidyl cyclase helical domain in catalytically active dimer formation. *Proc. Natl. Acad. Sci. USA* **2017**, *114*, 9821–9828. [CrossRef]
88. McIntosh, B.E.; Hogenesch, J.B.; Bradfield, C.A. Mammalian Per-Arnt-Sim proteins in environmental adaptation. *Ann. Rev. Physiol.* **2010**, *72*, 625–645. [CrossRef] [PubMed]
89. Henry, J.T.; Crosson, S. Ligand-binding PAS domains in a genomic, cellular, and structural context. *Ann. Rev. Microbiol.* **2011**, *65*, 261–286. [CrossRef]
90. Oka, Y.; Matsushita, T.; Mochizuki, N.; Quail, P.H.; Nagatani, A. Mutant screen distinguishes between residues necessary for light-signal perception and signal transfer by phytochrome B. *PLoS Genet.* **2008**, *4*, e1000158. [CrossRef]
91. Pellequer, J.L.; Wager-Smith, K.A.; Kay, S.A.; Getzoff, E.D. Photoactive yellow protein: A structural prototype for the three-dimensional fold of the PAS domain superfamily. *Proc. Natl. Acad. Sci. USA* **1998**, *95*, 5884–5890. [CrossRef]
92. Cheng, Z.; Yamamoto, H.; Bauer, C.E. Cobalamin's (Vitamin B12) Surprising Function as a Photoreceptor. *Trends Biochemi. Sci.* **2016**, *41*, 647–650. [CrossRef]
93. Drennan, C.L.; Huang, S.; Drummond, J.T.; Matthews, R.G.; Lidwig, M.L. How a protein binds B12: A 3.0 A X-ray structure of B12-binding domains of methionine synthase. *Science* **1994**, *266*, 1669–1674. [CrossRef] [PubMed]
94. Ludwig, M.L.; Matthews, R.G. Structure-based perspectives on B12-dependent enzymes. *Annu. Rev. Biochem.* **1997**, *66*, 269–313. [CrossRef] [PubMed]
95. Padmanabhan, S.; Jost, M.; Drennan, C.L.; Elías-Arnanz, M. A new facet of vitamin B12: Gene regulation by cobalamin-based photoreceptors. *Ann. Rev. Biochem.* **2017**, *86*, 485–514. [CrossRef] [PubMed]
96. Ortiz-Guerrero, J.M.; Polanco, M.C.; Murillo, F.J.; Padmanabhan, S.; Elias-Arnanz, M. Light-dependent gene regulation by a coenzyme B12-based photoreceptor. *Proc. Natl. Acad. Sci. USA* **2011**, *108*, 7565–7570. [CrossRef] [PubMed]
97. Jost, M.; Simpson, J.H.; Drennan, C.L. The transcription factor CarH safeguards use of adenosylcobalamin as a light sensor by altering the photolysis products. *Biochemistry* **2015**, *54*, 3231–3234. [CrossRef] [PubMed]
98. Kutta, R.J.; Hardman, S.J.; Johannissen, L.O.; Bellina, B.; Messiha, H.L.; Ortiz-Guerrero, J.M.; Elias-Arnanz, M.; Padmanabhan, S.; Barran, P.; Scrutton, N.S.; et al. The photochemical mechanism of a B12-dependent photoreceptor protein. *Nat. Commun.* **2015**, *6*, 7907. [CrossRef] [PubMed]
99. Jost, M.; Fernandez-Zapata, J.; Polanco, M.C.; Ortiz-Guerrero, J.M.; Chen, P.Y.; Kang, G.; Padmanabhan, S.; Elias-Arnanz, M.; Drennan, C.L. Structural basis for gene regulation by a B12-dependent photoreceptor. *Nature* **2015**, *526*, 536–541. [CrossRef]
100. Roth, J.R.; Lawrence, J.G.; Bobik, T.A. Cobalamin (coenzymeB12): Synthesis and biological significance. *Annu. Rev. Microbiol.* **1996**, *50*, 137–181. [CrossRef] [PubMed]
101. Cerutti, P.; Guroff, G. Enzymatic formation of phenylpyruvic acid in *Pseudomonas* sp. (ATCC 11299A) and its regulation. *J. Biol. Chem.* **1965**, *240*, 3034–3038. [PubMed]
102. Cotton, R.G.; Gibson, F. The biosynthesis of phenylalanine and tyrosine; enzymes converting chorismic acid into prephenic acid and their relationships to prephenate dehydratase and prephenate dehydrogenase. *Biochim. Biophys. Acta* **1965**, *100*, 76–88. [CrossRef]
103. Schmidt, J.C.; Zalkin, H. Chorismate mutase-prephenate dehydratase. Partial purification and properties of the enzyme from *Salmonella typhimurium*. *Biochemistry* **1969**, *8*, 174–181.

104. Kaushik, M.S.; Srivastava, M.; Mishra, A.K. Iron homeostasis in cyanobacteria. In *Cyanobacteria: From Basic Science toApplications*, 1st ed.; Mishra, A.K., Tiwari, D.N., Rai, A.N., Eds.; Academic Press: Cambridge, MA, USA, 2018; Volume 1, pp. 173–191.
105. Harrison, D.H.; Runquist, J.A.; Holub, A.; Miziorko, H.M. The crystal structure of phosphoribulokinase from *Rhodobacter sphaeroides* reveals a fold similar to that of adenylate kinase. *Biochemistry* **1998**, *37*, 5074–5085. [CrossRef] [PubMed]
106. Neuhard, J.; Tarpo, L. Location of the udk gene on the physical map of *Escherichia coli*. *J. Bacteriol.* **1993**, *175*, 5742–5743. [CrossRef] [PubMed]
107. Song, W.J.; Jackowski, S. Cloning, sequencing, and expression of the pantothenate kinase (coaA) gene of *Escherichia coli*. *J. Bacteriol.* **1992**, *174*, 6411–6417. [CrossRef] [PubMed]
108. Jergic, S.; Ozawa, K.; Williams, N.K.; Su, X.-C.; Scott, D.D.; Hamdan, S.M.; Crowther, J.A.; Otting, G.; Dixon, N.E. The unstructured C-terminus of the q subunit of *Escherichia coli* DNA polymerase III holoenzyme is the site of interaction with the a subunit. *Nucleic Acid Res.* **2007**, *35*, 2813–2824. [CrossRef] [PubMed]
109. Snyder, A.K.; Williams, C.R.; Johnson, A.; O'Donnell, M.; Bloom, L.B. Mechanism of loading the *Escherichia coli* DNA polymerase III sliding clamp. II. Uncoupling the b and DNA binding activities of the g complex. *J. Biol. Chem.* **2004**, *279*, 4386–4393. [CrossRef] [PubMed]
110. Jeruzalmi, D.; O'Donnell, M.; Kuriyan, J. Crystal Structure of the Processivity Clamp Loader Gamma (γ) Complex of *E. coli* DNA Polymerase III. *Cell* **2001**, *106*, 429–441. [CrossRef]
111. Tsuchihashi, Z.; Kornberg, A. Translational frameshifting generates the g subunit of DNA polymerase III holoenzyme. *Proc. Natl. Acad. Sci. USA* **1990**, *87*, 2516–2520. [CrossRef]
112. Cullman, G.; Fien, K.; Kobayashi, R.; Stillman, B. Characterization of the five replication factor C genes of *Saccharomyces cerevisiae*. *Mol. Cell. Biol.* **1995**, *15*, 4661–4671. [CrossRef]
113. O'Donnell, M.; Onrust, R.; Dean, F.B.; Chen, M.; Hurwitz, J. Homology in accessory proteins of replicative polymerases—*E. coli* to humans. *Nucleic Acids Res.* **1993**, *21*, 1–3. [CrossRef]
114. Munro, A.W.; McLean, K.J.; Grant, J.L.; Makris, T.M. Structure and function of the cytochrome P450 peroxygenase enzymes. *Biochem. Soc. Trans.* **2018**, *46*, 183–196. [CrossRef] [PubMed]
115. Li, L.; Chang, Z.; Pan, Z.; Fu, Z.-Q.; Wang, X. Modes of heme binding and substrate access for cytochrome P450 CYP74A revealed by crystal structures of allene oxide synthase. *Proc. Natl. Acad. Sci. USA* **2008**, *105*, 13883–13888. [CrossRef] [PubMed]
116. Fujishiro, T.; Shoji, O.; Nagano, S.; Sugimoto, H.; Shiro, Y.; Watanabe, Y. Crystal structure of H_2O_2-dependent cytochrome P450 SPα with its bound fatty acid substrate: Insight into the regioselective hydroxylation of fatty acids at the α position. *J. Biol. Chem.* **2011**, *286*, 29941–29950. [CrossRef] [PubMed]
117. Pao, G.M.; Saier, M.H. Response regulators of bacterial signal transduction systems: Selective domain shuffling during evolution. *J. Mol. Evol.* **1995**, *40*, 136–154. [CrossRef] [PubMed]
118. Galperin, M.Y.; Nikolskaya, A.N. Identification of sensory and signal-transducing domains in two-component signaling systems. *Met. Enzymol.* **2007**, *422*, 47–74.
119. Orth, P.; Schnappinger, D.; Hillen, W.; Saenger, W.; Hinrichs, W. Structural basis of gene regulation by the tetracycline inducible Tet repressor-operator system. *Nat. Struct. Biol.* **2000**, *7*, 215–219. [PubMed]
120. Ramos, J.L.; Martínez-Bueno, M.; Molina-Henares, A.J.; Terán, W.; Watanabe, K.; Zhang, X.; Gallegos, M.T.; Brennan, R.; Tobes, R. The TetR family of transcriptional repressors. *Microbiol. Mol. Biol. Rev.* **2005**, *69*, 326–356. [CrossRef]
121. Martinez-Bueno, M.; Molina-Henares, A.J.; Pareja, E.; Ramos, J.L.; Tobes, R. BacT regulators: A database of transcriptional regulators in bacteria and archaea. *Bioinformatics* **2004**, *20*, 2787–2791. [CrossRef]
122. Li, M.; Gu, R.; Su, C.C.; Routh, M.D.; Harris, K.C.; Jewell, E.S.; McDermott, G.; Edward, W.Y. Crystal structure of the transcriptional regulator AcrR from *Escherichia coli*. *J. Mol. Biol.* **2007**, *374*, 591–603. [CrossRef]
123. Su, C.C.; Rutherford, D.J.; Yu, E.W. Characterization of the multidrug efflux regulator AcrR from *Escherichia coli*. *Biochem. Biophys. Res. Commun.* **2007**, *361*, 85–90. [CrossRef]
124. Bruner, S.D.; Norman, D.P.; Verdine, G.L. Structural basis for recognition and repair of the endogenous mutagen 8-oxoguanine in DNA. *Nature* **2000**, *403*, 859. [CrossRef] [PubMed]
125. Nash, H.M.; Bruner, S.D.; Schärer, O.D.; Kawate, T.; Addona, T.A.; Spooner, E.; Lane, W.S.; Verdine, G.L. Cloning of a yeast 8-oxoguanine DNA glycosylase reveals the existence of a base-excision DNA-repair protein superfamily. *Curr. Biol.* **1996**, *6*, 968–980. [CrossRef]

126. Friedhoff, P.; Gimadutdinow, O.; Pingoud, A. Identification of catalytically relevant amino acids of the extracellular *Serratia marcescens* endonuclease by alignment-guided mutagenesis. *Nucleic Acid Res.* **1994**, *22*, 3280–3287. [CrossRef] [PubMed]
127. Miller, M.D.; Tanner, J.; Alpaugh, M.; Benedik, M.J.; Krause, K.L. 2. 1 Å structure of Serratia endonuclease suggests a mechanism for binding to double-stranded DNA. *Nat. Struct. Mol. Biol.* **1994**, *1*, 461. [CrossRef]
128. Schleif, R. AraC protein, regulation of the l-arabinose operon in *Escherichia coli*, and the light switch mechanism of AraC action. *FEMS Microbiol. Rev.* **2010**, *34*, 779–796. [CrossRef]
129. Rodgers, M.; Schleif, R. DNA tape measurements of AraC. *Nucleic Acids Res.* **2008**, *36*, 404–410. [CrossRef]
130. Ollis, D.L.; Cheah, E.; Cygler, M.; Dijkstra, B.; Frolow, F.; Franken, S.M.; Harel, M.; Remington, S.J.; Silman, I.; Schrag, J.; et al. The α/β hydrolase fold. *Protein Eng.* **1992**, *5*, 197–211. [CrossRef] [PubMed]
131. Nardini, M.; Dijkstra, B.W. α/β hydrolase fold enzymes: The family keeps growing. *Curr. Opin. Struct. Biol.* **1999**, *9*, 732–737. [CrossRef]
132. Carr, P.D.; Ollis, D.L. α/β hydrolase fold: An update. *Protein Pept. Lett.* **2009**, *16*, 1137–1148.
133. Goodacre, N.F.; Gerloff, D.L.; Uetz, P. Protein domains of unknown function are essential in bacteria. *MBio* **2014**, *5*, e00744-13. [CrossRef]
134. Mandiyan, V.; Andreev, J.; Schlessinger, J.; Hubbard, S.R. Crystal structure of the ARF-GAP domain and ankyrin repeats of PYK2-associated protein β. *EMBO J.* **1999**, *18*, 6890–6898. [CrossRef] [PubMed]
135. Lux, S.E.; John, K.M.; Bennett, V. Analysis of cDNA for human erythrocyte ankyrin indicates a repeated structure with homology to tissue-differentiation and cell-cycle control proteins. *Nature* **1990**, *344*, 36. [CrossRef] [PubMed]
136. Bork, P. Hundreds of ankyrin-like repeats in functionally diverse proteins: Mobile modules that cross phyla horizontally? *Proteins* **1993**, *17*, 363–374. [CrossRef] [PubMed]
137. Bork, P.; Doolittle, R.F. Proposed acquisition of an animal protein domain by bacteria. *Proc. Natl. Acad. Sci. USA* **1992**, *89*, 8990–8994. [CrossRef] [PubMed]
138. Davis, L.H.; Otto, E.; Bennett, V. Specific 33-residue repeat (s) of erythrocyte ankyrin associate with the anion exchanger. *J. Biol. Chem.* **1991**, *266*, 11163–11169.
139. Hall, A. Rho GTPases and the actin cytoskeleton. *Science* **1998**, *279*, 509–514. [CrossRef]
140. Crespo, P.; Schuebel, K.E.; Ostrom, A.A.; Gutkind, J.S.; Bustelo, X.R. Phosphotyrosine-dependent activation of Rac-1 GDP/GTP exchange by the vav proto-oncogene product. *Nature* **1997**, *385*, 169. [CrossRef] [PubMed]
141. Hart, M.J.; Eva, A.; Evans, T.; Aaronson, S.A.; Cerione, R.A. Catalysis of guanine nucleotide exchange on the CDC42Hs protein by the dbl oncogene product. *Nature* **1991**, *354*, 311. [CrossRef]
142. Takano, H.; Komuro, I.; Oka, T.; Shiojima, I.; Hiroi, Y.; Mizuno, T.; Yazaki, Y. The Rho family G proteins play a critical role in muscle differentiation. *Mol. Cell. Biol.* **1998**, *18*, 1580–1589. [CrossRef]
143. Moorman, J.P.; Luu, D.; Wickham, J.; Bobak, D.A.; Hahn, C.S. A balance of signaling by Rho family small GTPases RhoA, Rac1 and Cdc42 coordinates cytoskeletal morphology but not cell survival. *Oncogene* **1999**, *18*, 47. [CrossRef]
144. Kaibuchi, K.; Kuroda, S.; Amano, M. Regulation of the cytoskeleton and cell adhesion by the Rho family GTPases in mammalian cells. *Ann. Rev. Biochem.* **1999**, *68*, 459–486. [CrossRef] [PubMed]
145. Chimini, G.; Chavrier, P. Function of Rho family proteins in actin dynamics during phagocytosis and engulfment. *Nat. Cell Biol.* **2000**, *2*, 191. [CrossRef] [PubMed]
146. Prokopenko, S.N.; Saint, R.; Bellen, H.J. Untying the Gordian knot of cytokinesis: Role of small G proteins and their regulators. *J. Cell Biol.* **2000**, *148*, 843–848. [CrossRef] [PubMed]
147. Somlyo, A.P.; Somlyo, A.V. Signal transduction by G-proteins, rho-kinase and protein phosphatase to smooth muscle and non-muscle myosin II. *J. Physiol.* **2000**, *522*, 177–185. [CrossRef] [PubMed]
148. Uehata, M.; Ishizaki, T.; Satoh, H.; Ono, T.; Kawahara, T.; Morishita, T.; Tamakawa, H.; Yamagami, K.; Inui, J.; Maekawa, M.; et al. Calcium sensitization of smooth muscle mediated by a Rho-associated protein kinase in hypertension. *Nature* **1997**, *389*, 990. [CrossRef]
149. Zohn, I.M.; Campbell, S.L.; Khosravi-Far, R.; Rossman, K.L.; Der, C.J. Rho family proteins and Ras transformation: The RHOad less traveled gets congested. *Oncogene* **1998**, *17*, 1415. [CrossRef] [PubMed]
150. Nishida, K.; Kaziro, Y.; Satoh, T. Association of the proto-oncogene product Dbl with G protein βγ subunits. *FEBS Lett.* **1999**, *459*, 186–190. [CrossRef]
151. Cerione, R.A.; Zheng, Y. The Dbl family of oncogenes. *Curr. Opin. Cell Biol.* **1996**, *8*, 216–222. [CrossRef]

152. Longenecker, K.L.; Lewis, M.E.; Chikumi, H.; Gutkind, J.S.; Derewenda, Z.S. Structure of the RGS-like domain from PDZ-RhoGEF: Linking heterotrimeric g protein-coupled signaling to Rho GTPases. *Structure* **2001**, *9*, 559–569. [CrossRef]
153. Ponting, C.P. Evidence for PDZ domains in bacteria, yeast, and plants. *Protein Sci.* **1997**, *6*, 464–468. [CrossRef]
154. Ponting, C.P.; Phillips, C.; Davies, K.E.; Blake, D.J. PDZ domains: Targeting signaling molecules to sub-membranous sites. *Bioessays* **1997**, *19*, 469–479. [CrossRef] [PubMed]
155. Grootjans, J.J.; Zimmermann, P.; Reekmans, G.; Smets, A.; Degeest, G.; Dürr, J.; David, G. Syntenin, a PDZ protein that binds syndecan cytoplasmic domains. *Proc. Natl. Acad. Sci. USA* **1997**, *94*, 13683–13688. [CrossRef] [PubMed]
156. Lohr, D.; Venkov, P.; Zlatanova, J. Transcriptional regulation in the yeast GAL gene family: A complex genetic network. *FASEB J.* **1995**, *9*, 777–787. [CrossRef] [PubMed]
157. Marmorstein, R.; Carey, M.; Ptashne, M.; Harrison, S.C. DNA recognition by GAL4: Structure of a protein-DNA complex. *Nature* **1992**, *356*, 408. [CrossRef] [PubMed]
158. Pan, T.; Coleman, J.E. GAL4 transcription factor is not a "zinc finger" but forms a Zn (II) 2Cys6 binuclear cluster. *Proc. Natl. Acad. Sci. USA* **1990**, *87*, 2077–2081. [CrossRef] [PubMed]
159. Todd, R.B.; Andrianopoulos, A. Evolution of a fungal regulatory gene family: The Zn (II) 2Cys6 binuclear cluster DNA binding motif. *Fungal Genet. Biol.* **1997**, *21*, 388–405. [CrossRef] [PubMed]
160. Pathak, G.P.; Vrana, J.D.; Tucker, C.L. Optogenetic control of cell function using engineered photoreceptors. *Bio. Cell* **2013**, *105*, 59–72. [CrossRef]
161. Fiedler, B.; Broc, D.; Schubert, H.; Rediger, A.; Börner, T.; Wilde, A. Involvement of cyanobacterial phytochromes in growth under different light qualities and quantities. *Photochem. Photobiol.* **2004**, *79*, 551–555. [CrossRef] [PubMed]
162. Fischer, A.J.; Rockwell, N.C.; Jang, A.Y.; Ernst, L.A.; Waggoner, A.S.; Duan, Y.; Lei, H.; Lagarias, J.C. Multiple roles of aconserved GAF domain tyrosine residue in cyanobacterial and plant phytochromes. *Biochemistry* **2005**, *44*, 15203–15215. [CrossRef]
163. Kehoe, D.M.; Grossman, A.R. New classes of mutants in complementary chromatic adaptation provide evidence for a novel four-step phosphorelay system. *J. Bacteriol.* **1997**, *179*, 3914–3921. [CrossRef]
164. Moon, Y.J.; Park, Y.M.; Chung, Y.H.; Choi, J.S. Calcium is involved in photomovement of cyanobacterium *Synechocystis* sp. PCC 6803. *Photochem. Photobiol.* **2004**, *79*, 114–119. [CrossRef]
165. Hong, K.P.; Spitzer, N.C.; Nicol, X. Improved molecular toolkit for cAMP studies in live cells. *BMC Res. Notes* **2011**, *4*, 241. [CrossRef] [PubMed]
166. Yoshihara, S.; Katayama, M.; Geng, X.; Ikeuchi, M. Cyanobacterial phytochrome-like PixJ1 holoprotein shows novel reversible photoconversion between blue and green-absorbing forms. *Plant Cell Physiol.* **2004**, *45*, 1729–1737. [CrossRef] [PubMed]
167. Yoshihara, S.; Shimada, T.; Matsuoka, D.; Zikihara, K.; Kohchi, T.; Tokutomi, S. Reconstitution of blue-green reversible photoconversion of a cyanobacterial photoreceptor, PixJ1, in phycocyanobilin-producing *Escherichia coli*. *Biochemistry* **2006**, *45*, 3775–3784. [CrossRef]
168. Kehoe, D.M.; Grossman, A.R. Similarity of achromatic adaptation sensor to phytochrome and ethylene receptors. *Science* **1996**, *273*, 1409–1412. [CrossRef] [PubMed]
169. Schröder-Lang, S.; Schwärzel, M.; Seifert, R.; Strünker, T.; Kateriya, S.; Looser, J.; Watanabe, M.; Kaupp, U.B.; Hegemann, P.; Nagel, G. Fast manipulation of cellular cAMP level by light in vivo. *Nat. Methods* **2007**, *4*, 39–42. [CrossRef] [PubMed]
170. Nagahama, T.; Suzuki, T.; Yoshikawa, S.; Iseki, M. Functional transplant of photoactivated adenylyl cyclase (PAC) into Aplysia sensory neurons. *Neurosci. Res.* **2007**, *59*, 81–88. [CrossRef]
171. Weissenberger, S.; Schultheis, C.; Liewald, J.F.; Erbguth, K.; Nagel, G.; Gottschalk, A. PACα—An optogenetic tool for in vivo manipulation of cellular cAMP levels, neurotransmitter release and behavior in *Caenorhabditis elegans*. *J. Neurochem.* **2011**, *116*, 616–625. [CrossRef]
172. Cheng, Z.; Li, K.; Hammad, L.A.; Karty, J.A.; Bauer, C.E. Vitamin B12 regulates photosystem gene expression via the CrtJantirepressorAerR in *Rhodobactercapsulatus*. *Mol. Microbiol.* **2014**, *91*, 649–664. [CrossRef]
173. Zoltowski, B.D.; Vaccaro, B.; Crane, B.R. Mechanism-based tuning of a LOV domain photoreceptor. *Nat. Chem. Biol* **2009**, *5*, 827–834. [CrossRef]

174. Stephan, E.; Granzin, J.; Circolone, F.; Stadler, A.; Krauss, U.; Drepper, T.; Svensson, V.; Knieps-Grünhagen, E.; Wirtz, A.; Cousin, A.; et al. Structure and function of a short LOV protein from the marine phototrophic bacterium *Dinoroseobactershibae*. *BMC Microbiol.* **2015**, *15*, 30.
175. Pillai, V.R. Photoremovable protecting groups in organic synthesis. *Synthesis* **1980**, *1980*, 1–26. [CrossRef]
176. Cardin, J.A.; Carlén, M.; Meletis, K.; Knoblich, U.; Zhang, F.; Deisseroth, K.; Tsai, L.H.; Moore, C.I. Targeted optogenetic stimulation and recording of neurons in vivo using cell-type-specific expression of Channelrhodopsin-2. *Nat. Protoc.* **2010**, *5*, 247. [CrossRef] [PubMed]
177. Levskaya, A.; Weiner, O.D.; Lim, W.A.; Voigt, C.A. Spatiotemporal control of cell signaling using a light-switchable protein interaction. *Nature* **2009**, *461*, 997. [CrossRef] [PubMed]
178. Kennedy, M.J.; Hughes, R.M.; Peteya, L.A.; Schwartz, J.W.; Ehlers, M.; Tucker, C.L. Rapid blue-light–mediated induction of protein interactions in living cells. *Nat. Methods* **2010**, *7*, 973. [CrossRef] [PubMed]

Article

Ion Channel Properties of a Cation Channelrhodopsin, *Gt*_CCR4

Shunta Shigemura [1], Shoko Hososhima [1], Hideki Kandori [1,2] and Satoshi P. Tsunoda [1,3,*]

1 Department of Life Science and Applied Chemistry, Nagoya Institute of Technology, Showa-ku, Nagoya 466-8555, Japan
2 OptoBio Technology Research Center, Nagoya Institute of Technology, Showa-ku, Nagoya 466-8555, Japan
3 PRESTO, Japan Science and Technology Agency, 4-1-8 Honcho, Kawaguchi, Saitama 332-0012, Japan
* Correspondence: tsunoda.satoshi@nitech.ac.jp; Tel.: +81-52-735-5218

Received: 23 July 2019; Accepted: 18 August 2019; Published: 21 August 2019

Abstract: We previously reported a cation channelrhodopsin, *Gt*_CCR4, which is one of the 44 types of microbial rhodopsins from a cryptophyte flagellate, *Guillardia theta*. Due to the modest homology of amino acid sequences with a chlorophyte channelrhodopsin such as *Cr*_ChR2 from *Chlamydomonas reinhardtii*, it has been proposed that a family of cryptophyte channelrhodopsin, including *Gt*_CCR4, has a distinct molecular mechanism for channel gating and ion permeation. In this study, we compared the photocurrent properties, cation selectivity and kinetics between well-known *Cr*_ChR2 and *Gt*_CCR4 by a conventional path clamp method. Large and stable light-induced cation conduction by *Gt*_CCR4 at the maximum absorbing wavelength (530 nm) was observed with only small inactivation (15%), whereas the photocurrent of *Cr*_ChR2 exhibited significant inactivation (50%) and desensitization. The light sensitivity of *Gt*_CCR4 was higher (EC_{50} = 0.13 mW/mm^2) than that of *Cr*_ChR2 (EC_{50} = 0.80 mW/mm^2) while the channel open life time (photocycle speed) was in the same range as that of *Cr*_ChR2 (25~30 ms for *Gt*_CCR4 and 10~15 ms for *Cr*_ChR2). This observation implies that *Gt*_CCR4 enables optical neuronal spiking with weak light in high temporal resolution when applied in neuroscience. Furthermore, we demonstrated high Na^+ selectivity of *Gt*_CCR4 in which the selectivity ratio for Na^+ was 37-fold larger than that for *Cr*_ChR2, which primarily conducts H^+. On the other hand, *Gt*_CCR4 conducted almost no H^+ and no Ca^{2+} under physiological conditions. These results suggest that ion selectivity in *Gt*_CCR4 is distinct from that in *Cr*_ChR2. In addition, a unique red-absorbing and stable intermediate in the photocycle was observed, indicating a photochromic property of *Gt*_CCR4.

Keywords: microbial rhodopsin; channelrhodopsin; electrophysiology; optogenetics

1. Introduction

Microbial-type rhodopsins are made up of seven or eight transmembrane helices with a covalently bound all-*trans* retinal as the chromophore [1]. They are found in archaea, bacteria, eukaryota (such as fungi and algae) and viruses, and are physiologically responsible for energy production and the phototaxis reaction. Molecular functions of microbial rhodopsin involve ion transporters, sensors and light-regulated enzymes. As for ion-transporting rhodopsins, they are divided into ion-pumps and channels. Bacteriorhodopsin (BR) was the first identified outward directed proton-pumping rhodopsin [2]. The discovery of a Cl^- pump, an Na^+ pump and inward-directed proton pumps has been even until recently [3–6]. Structure-based and spectroscopic studies, when combined with electrophysiology and molecular dynamics studies, revealed the detailed molecular mechanism of bacteriorhodopsin and other pumps.

Channelrhodopsin-1 and -2 (*Cr*_ChR1 and *Cr*_ChR2) from *Chlamydomonas reinhardtii* were the first light-gated ion channels to be discovered [7,8]. These homologous proteins permeate cations in which the permeability ratio of H^+, Na^+, and K^+ is 10^6, 1, and 0.5, respectively. High-resolution X-ray structures revealed details of their molecular architecture and provided insight into their photoactivation and ion conduction [9,10].

Successful expression of *Cr*_ChR2 in neurons allowed the action potential to be manipulated by light, which opened up a new field of research, optogenetics [11,12]. A number of variant molecules have been engineered to improve the functionality of ChR, and homologous ChRs were then reported [13]. Color-tuning variants cover almost the entire visible range. *Cr*_ChR2 displays an action spectrum maximum at 470 nm [8]. ChR variants such as C1V1, which is the chimeric version of ChR1 from *Chlamydomonas reinhardtii* and *Volvox carteri*, or C1C2 (a green receiver) absorb light at around 530~545 nm [14–16]. Another red-shifted ChR, Chrimson from *Chlamydomonas noctigama*, exhibits an absorption maximum at 590 nm which allows reliable neuronal stimulation by light exceeds 600 nm [17]. On the other hand, *Ts*ChR or *Ps*ChR absorb a shorter wavelength, making it possible to excite neurons at 440 nm [18].

The lifetime of an open channel can be extended by mutations at C128 and D156 (DC pair) which form a hydrogen bond bridge in *Cr*_ChR2. Mutations at C128 to Thr, Ala and Ser slowed the kinetics of channel closing 200, 5000 and 10,000-fold respectively [19]. *Cr*_ChR2 D156C displayed an even stronger effect, namely higher light sensitivity and prolonged lifetime of the open channel, by as much as 30 min [20].

Converting ion selectivity is challenging. The *Cr*_ChR2 L132C mutant showed improved Ca^{2+} permeability [21]. The permeability ratio between H^+ and Na^+ could be modified by a replacement at E143 to A in Chrimson [22]. Anion channelrhodopsins were engineered or discovered from nature and they have been applied as neuronal silencing tools [23–25]. The crystal structures of anion channelrhodopsins revealed their unique features related to the channel gating mechanism [26–28].

A novel cation channelrhodopsin family was reported in 2016 and 2017 from *Guillardia theta*, namely *Gt*_CCR1-4 [29,30]. These cation channelrhodopsins (CCRs) from cryptophyte algae are more homologous to haloarchaeal rhodopsins, such as proton pumping bacteriorhodopsin, than to chlorophyte CCRs, including *Cr*_ChR2. Actually, *Gt*_CCRs conserve the characteristic amino-acid residues involved in unidirectional proton transfer, including the proton acceptor D85 and the proton donor D96 in bacteriorhodopsin (Table S1).

On the other hand, a characteristic glutamic acid in TM2 (E90 in *Cr*_ChR2) which is crucial for channel gating and ion selectivity, is not conserved in *Gt*_CCRs [23,31]. *Cr*_ChR2 possesses a so-called DC pair (C128 and D156 in *Cr*_ChR2), which is responsible for the channel life time [19,32,33]. This is not found in *Gt*_CCRs. Thus, overall sequence patterns separate these cryptophyte CCRs form chlorophyte channels. The molecular mechanisms such as channel gating mechanism and ion selectivity could be distinct in chlorophyte CCRs. Sineshchekov and coworkers already revealed that the retinal Schiff-base (SB) in *Gt*_CCR2 rapidly deprotonates to the D85 homolog, as in BR, upon photoisomerization [34]. Channel-opening requires deprotonation of the D96 homolog. We independently identified photocycle intermediates during the channel function of *Gt*_CCR4 from electrophysiological and flash photolysis experiments. The M-decay corresponds to channel-closing, implicating tight coupling between retinal dynamics and channel function. However, reprotonation of SB for channel closing was achieved by the direct return of a proton from the D85 homolog. Such proton transfers are not the case with *Cr*_ChR2. In *Cr*_ChR2, D156 in TM4 provides the proton [35]. We demonstrated, using an FTIR study, that the secondary structural change in the primary reaction was much smaller than in *Cr*_ChR2 [30]. These differences in the molecular mechanism place the cryptophyte CCR in a new family of channelrhodopsins, which we described as "DTD channelrhodopsins" or "BR-like cation channelrhodopsins" [29,30]. To further reveal the characteristics of these DTD channelrhodopsins, in this study we performed electrophysiological measurements in parallel with *Cr*_ChR2.

2. Materials and Methods

2.1. Expression Plasmids

A mammalian expression plasmid peGFP-*Gt*_CCR4 was described previously [30]. pVenus-N1-Chop2-315 was a kind gift from Prof. Yawo (The University of Tokyo) [12].

2.2. Cell Culture

The electrophysiological assays of *Gt*_CCR4 and *Cr*_ChR2 were performed on ND7/23 cells, which are hybrid cell lines derived from neonatal rat dorsal root ganglia neurons fused with mouse neuroblastoma [36]. ND7/23 cells were grown on a collagen-coated coverslip in Dulbecco's modified Eagle's medium (Wako, Osaka, Japan) supplemented with 2.0 μM all-*trans* retinal and 5% fetal bovine serum, and under a 5% CO_2 atmosphere at 37 °C. The expression plasmids were transiently transfected by using the FuGENE HD transfection Reagent (Promega, Fitchburg, WI, USA) according to the manufacturer's instructions. Electrophysiological recordings were then conducted 24–36 h after transfection. Successfully transfected cells were identified by eGFP or Venus fluorescence under a microscope prior to the measurements.

2.3. Electrophysiology

All experiments were carried out at room temperature (22 ± 2 °C). Photocurrents were recorded as previously described using an Axopatch 200B amplifier (Molecular Devices, Sunnyvale, CA, USA) under a whole-cell patch clamp configuration [12]. Data were filtered at 5 kHz and sampled at 20 kHz (Digdata1550, Molecular Devices, Sunnyvale, CA, USA) and stored in a computer (pClamp10.6, Molecular Devices). Pipette resistance was 3–6 MΩ. The standard internal pipette solution for the whole-cell voltage clamp contained (in mM) 120 KOH, 100 glutamate, 2.5 $MgCl_2$, 2.5 MgATP, 0.01 Alexa568, 50 HEPES, and 5 EGTA, and adjusted to pH 7.2. The standard extracellular solution for the whole-cell voltage clamp contained (in mM) 140 NaCl, 2 KCl, 2 $MgCl_2$, 2 $CaCl_2$, and 10 HEPES, and adjusted to pH 7.2. The ion selectivity internal pipette solution for the whole-cell voltage clamp contained (in mM) 1 NaCl, 1 KCl, 2 $CaCl_2$, 2 $MgCl_2$, 110 N-methyl D-glucamine, 10 CHES, and 10 EGTA, and adjusted to pH 9.0. The ion selectivity extracellular solution for the whole-cell voltage clamp contained (in mM) $Ex_{NMG,\,9.0}$ is 1 NaCl, 1 KCl, 2 $CaCl_2$, 2 $MgCl_2$, 140 N-methyl D-glucamine, and 10 CHES, and adjusted to pH 9.0. $Ex_{NMG,\,6.85}$ is 1 NaCl, 1 KCl, 2 $CaCl_2$, 2 $MgCl_2$, 140 N-methyl D-glucamine, and 10 MES, and adjusted to pH 6.85. $Ex_{NaCl,\,9.0}$ is 140 NaCl, 1 KCl, 2 $CaCl_2$, 2 $MgCl_2$, and 10 CHES, and adjusted to pH 9.0. $Ex_{KCl,\,9.0}$ is 1 NaCl, 140 KCl, 2 $CaCl_2$, 2 $MgCl_2$, and 10 CHES, and adjusted to pH 9.0. $Ex_{CsCl,\,9.0}$ is 1 NaCl, 1 KCl, 140 CsCl, 2 $CaCl_2$, 2 $MgCl_2$, and 10 CHES, and adjusted to pH 9.0. $Ex_{CaCl2,\,9.0}$ is 1 NaCl, 1 KCl, 70 $CaCl_2$, 2 $MgCl_2$, and 10 CHES, and adjusted to pH 9.0. $Ex_{MgCl2,\,9.0}$ is 1 NaCl, 1 KCl, 2 $CaCl_2$, 70 $MgCl_2$, and 10 CHES, and adjusted to pH 9.0. All solutions of pH were adjusted with N-methyl D-glucamine or HCl. The liquid junction potential was calculated and compensated by pClamp 10.6 software. Time constants were determined by a single exponential fit unless noted.

2.4. Optics

For whole-cell voltage clamp, irradiation at 470 or 530 or 590 nm was carried out using WheeLED and collimated LED (parts No. WLS-LED-0530-03 or LCS-0530-03-22, WLS-LED-0590-03 Mightex, Toronto, ON, Canada) or an ND YAG flash laser, Mini lite at 532 nm (Continuum, San Jose, CA, USA) controlled by computer software (pCLAMP10.6, Molecular Devices). Light power was measured directly by an objective lens of a microscope by a power meter (LP1, Sanwa Electric Instruments Co., Ltd., Tokyo, Japan).

2.5. Confocal Images

Cell images shown in Figure 1A,B were observed by a Nikon A1 LFOV through an objective lens, Apo 60x Oil λS DIC N2.

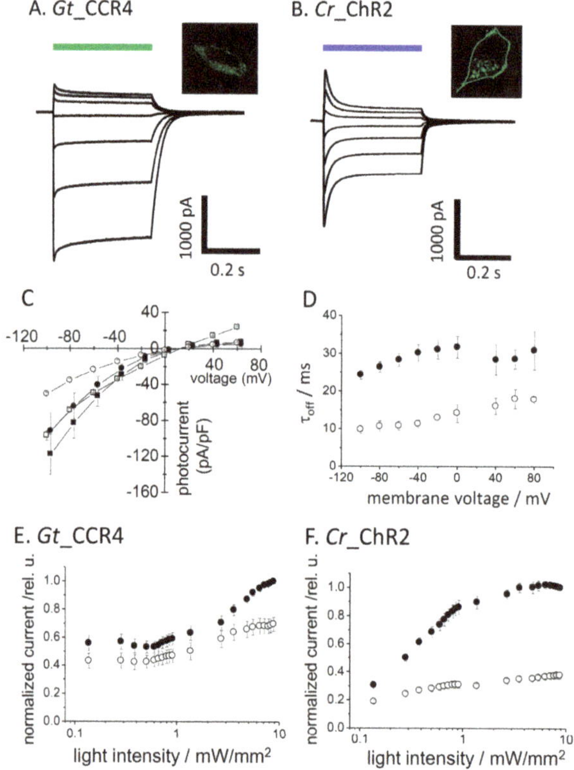

Figure 1. Basic properties of *Gt*_CCR4 and *Cr*_ChR2. (**A**,**B**) Each cation channelrhodopsin expressed in ND7/23 cells was stimulated by green (530 nm) or blue (470 nm) LED light (6.8 mW/mm^2). Standard solutions were used. Membrane potentials were clamped from −60 mV to +60 mV in +20 mV steps. Fluorescence images were taken using a confocal microscope. (**C**) Current-voltage relationship (I-V plot) of *Gt*_CCR4 (filled symbol) and *Cr*_ChR2 (empty symbol). Current peak component (square) and steady state amplitude (circle) of two channels are depicted. (**D**) Current-decay kinetics of *Gt*_CCR4 and *Cr*_ChR2. τ_{off} is plotted as a function of membrane voltage. Filled circle; *Gt*_CCR4, empty circle; *Cr*_ChR2. (**E**,**F**) Light power dependency of photocurrents from *Gt*_CCR4 and *Cr*_ChR2 at −60 mV. Each channel was stimulated by 530 nm (*Gt*_CCR4) and 470 nm (*Cr*_ChR2). Photocurrent values are normalized. Current peak component (filled circle) and steady state amplitude (empty circle) are depicted. (n = 4–8 cells).

2.6. Statistical Analysis

All data in the text and figures are expressed as mean ± SEM.

3. Results

3.1. Basic Characterization of Photocurrent

We transiently expressed *Gt*_CCR4 and *Cr*_ChR2 in ND7/23 cells by a conventional transfection method (FuGENE). Expression of these channels was visualized by tagged-GFP or Venus fluorescence. Strong membrane expression was confirmed for both channels, although cytoplasmic GFP was observed in *Gt*_CCR4-expressing cells (Figure 1A microscope images). We illuminated 530 nm light to induce a photocurrent for *Gt*_CCR4 and 470 nm light for *Cr*_ChR2 at the same light intensity (6.8 mW/mm^2). A large photocurrent was recorded from the *Gt*_CCR4-expressing cells, reproducing previous studies (Figure 1A, left) [30]. Current amplitude reached −2 nA at −60 mV. The current showed an initial peak component (I_p) which decayed slightly into a steady state level (I_{ss}). However, amplitude of the steady state still retained about 80% of the transient peak component. The photocurrent from *Cr*_ChR2 showed a large peak component which reached about −2 nA at −60 mV (Figure 1A, right). The current decayed by 50% of the initial peak, suggesting that *Cr*_ChR2 exhibits a markedly large inactivation compared to *Gt*_CCR4. Figure 1C depicts the current-voltage relationship of photocurrent from *Gt*_CCR4 and *Cr*_ChR2. Both peak component (I_p) and steady state current (I_{ss}) are plotted. The shape of the I-V plot from *Gt*_CCR4 indicates strong inward-rectification. I_{ss} of *Cr*_ChR2 displayed similarly inward-rectification, while I_p was weakly rectified, in which a markedly outward-directed current was observed at positive membrane voltages. For both I_p and I_{ss}, *Gt*_CCR4 showed a larger current density (pA/pF) than *Cr*_ChR2. For example, the photocurrent (I_{ss}) of *Gt*_CCR4 at −100 mV exceeded −80 pA/pF, while that of *Cr*_ChR2 was only about −40 pA/pF.

Kinetics in the photocurrent decay after shutting off the light is shown in Figure 1D. The time constant of *Gt*_CCR4 is about 25–35 ms under a membrane voltage between −100 and 80 mV, while *Cr*_ChR2 showed faster kinetics by about 10–20 ms. Next, we compared light sensitivity in the photocurrent of two channels (Figure 1E,F). The photocurrent amplitude from *Cr*_ChR2 grows as a typical sigmoidal curve for both the initial peak and the steady state components (Figure 1F). EC_{50} was determined as 0.8 mW/mm^2 for I_p and 0.35 mW/mm^2 for I_{ss} under the conditions tested. On the other hand, the *Gt*_CCR4 current showed unique growth in terms of power dependency with two apparent phases in which the current first saturated at 0.1 mW/mm^2 at about 50% of full activation, followed by the second phase of growth from 1 to 10 mW/mm^2 (Figure 1E). The EC_{50} was determined as 0.13 mW/mm^2 for I_p and 0.18 mW/mm^2 for I_{ss}. These results indicate that *Gt*_CCR4 is more sensitive to light with respect to channel activation.

3.2. Ion Selectivity

It was already reported that *Gt*_CCRs are H^+- and Na^+-permeable cation channels [29,30]. We here investigated the cation selectivity of *Gt*_CCR4 in more detail relative to *Cr*_ChR2. Ionic conditions of the extracellular solution were systematically exchanged with various cations including Na^+, K^+, Cs^+, Ca^{2+}, Mg^{2+} and NMG. In the presence of NMG, one can assume H^+ as the permeated ion. Figure 2A,B show the I-V plot of *Gt*_CCR4 and *Cr*_ChR2 under several ionic conditions. Obviously, the reversal potential shift in Na^+ solution is larger in *Gt*_CCR4 than in *Cr*_ChR2, suggesting that *Gt*_CCR4 is more permeable to Na^+ than *Cr*_ChR2. Notably, I-V plots of *Gt*_CCR4 at pH 6.85 and 9.0 in NMG are almost identical (Figure 2A) whereas a large shift of reversal potential was observed in *Cr*_ChR2 under the same conditions (Figure 2B). These results indicate that *Gt*_CCR4 has less H^+ selectivity than *Cr*_ChR2. The photocurrent amplitude of *Gt*_CCR4 and *Cr*_ChR2 at −60 mV under each condition is summarized in Figure 2C,D. The current amplitude of *Gt*_CCR4 was significantly larger in the presence of Na^+ and K^+ close to −100 pA/pF and in the presence of Cs^+ at about −40 pA/pF. In contrast, only a negligible current was observed at low pH, or in the presence of Ca^{2+} or Mg^{2+}. This supports the notion that *Gt*_CCR4 is more of a monovalent metal cation selective channel. In addition, we also tested measurements under a competitive environment in which both Na^+ and Ca^{2+} were both added to bath solutions (Figure S1). Interestingly, photocurrents by Gt_CCR4 were suppressed at a higher Ca^{2+} concentration (40 mM), suggesting that Na^+ flow is blocked by Ca^{2+}. In contrast, such a large difference in current amplitude was not observed under various conditions in the photocurrent from *Cr*_ChR2 (Figure 2D). This is due to low ion selectivity in *Cr*_ChR2 as was reported previously. To assume the selectivity ratio, the reversal potential shift from the condition with NMG at pH 9.0 is depicted in Figure 2E. The shifts (ΔE_{rev}) in *Gt*_CCR4 are larger than in *Cr*_ChR2 for Na^+, K^+, and Cs^+ indicating that *Gt*_CCR4 is selective for monovalent cations but less selective for H^+. In the initial study of *Cr*_ChR2, the permeability ratio was calculated based on the current amplitude [8]. The table in Figure 2F summarizes the ratio both for *Gt*_CCR4 and *Cr*_ChR2. H^+ permeability for *Cr*_ChR2 is 0.77×10^5, close to the reported value (1×10^6), while that for *Gt*_CCR4 is 2.1×10^4, indicating about 37-fold less permeability for H^+ in *Gt*_CCR4.

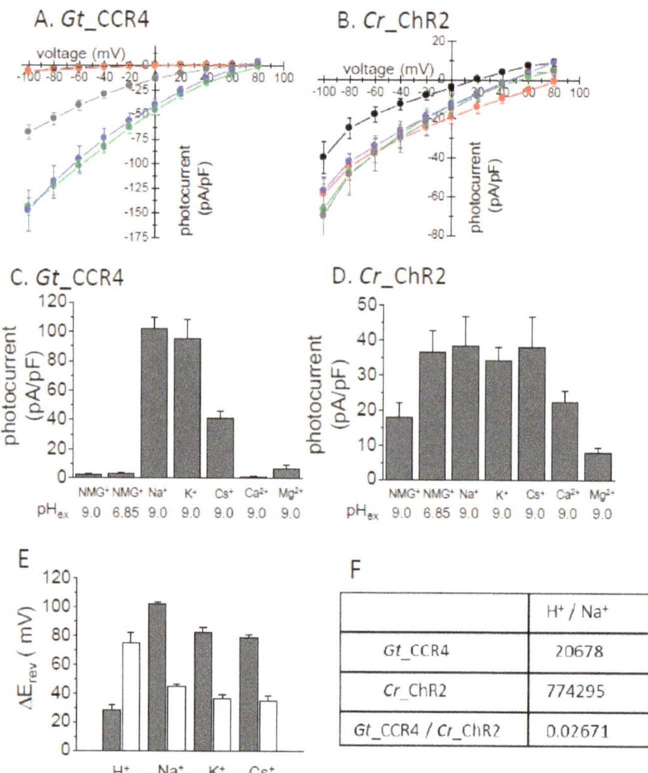

	H^+ / Na^+
*Gt*_CCR4	20678
*Cr*_ChR2	774295
*Gt*_CCR4 / *Cr*_ChR2	0.02671

Figure 2. Ion selectivity of *Gt*_CCR4 and *Cr*_ChR2. Each channelrhodopsin expressed in ND7/23 cells was stimulated by green and blue LED light. I-V plot of *Gt*_CCR4 (**A**) and *Cr*_ChR2 (**B**) are depicted. Steady-state current density (pA/pF) in 20 mV steps from −100 mV to +80 mV was plotted. The present liquid junction potential was considered. The pipette solution contained 110 mM NMG-Cl at pH 9.0, and the bath solution varied: black, NMG-Cl at pH 9.0; red, NMG-Cl at pH 6.85; green, NaCl at pH 9.0; blue, KCl at pH 9.0; grey, CsCl at pH 9.0. See "Materials and Methods" for details about the solutions. (**C**,**D**) Comparison of current density of *Gt*_CCR4 (**C**) and *Cr*_ChR2 (**D**) in the presence of various cations at −60 mV. (**E**) Reversal potential shift (ΔE_{rev}) for each condition for *Gt*_CCR4 and *Cr*_ChR2. E_{rev} was determined from the I-V plot shown in **A**,**B**. Each E_{rev} value was subtracted from the E_{rev} at NMG-Cl at pH 9.0. (**F**) Permeability ratio for H^+ and Na^+ in *Gt*_CCR4 and *Cr*_ChR2 as estimated from current value and ionic concentration. (n = 6–9 cells).

3.3. Flash Laser Electrophysiology

We then measured the photocurrent under a single-turnover condition with a flash laser as the light source (ND-YAG). As shown in Figure 3A,B, 5 ns light evoked a large inward-directed peak current at −60 mV for both *Gt*_CCR4- and *Cr*_ChR2-expressing cells. The current amplitude and direction are voltage-dependent for both channels, as was expected from recordings with LED. Current growth was fitted with a single exponential function (Figure 3C). The time constant of *Gt*_CCR4 seems to be independent of membrane voltage, while that of *Cr*_ChR2 slowed down slightly as voltage increased. Current decay was determined as about 20 ms for *Gt*_CCR4 and about 5–10 ms for *Cr*_ChR2, both of which are smaller than the value obtained from the measurement by LED light (Figure 1C), suggesting two distinct open states with different kinetics.

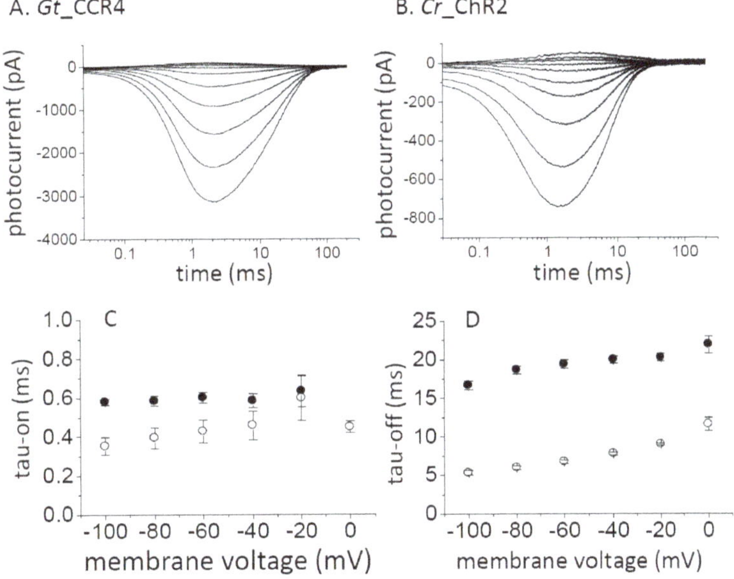

Figure 3. Flash laser stimulation. Standard solutions were used. Each channelrhodopsin in ND7/23 cells was stimulated by 5 ns by a green flash laser. Representative trace generated by *Gt*_CCR4 (**A**) and *Cr*_ChR2 (**B**) in 20 mV steps from −100 mV to +80 mV. (**C**) τ_{on}-voltage relationship from *Gt*_CCR4 (filled circle) and *Cr*_ChR2 (empty circle). Current rise was fitted by a single exponential function. The time constant was plotted. (**D**) τ_{off}-voltage relationship from *Gt*_CCR4 (filled circle) and *Cr*_ChR2 (empty circle). Current decay was fitted by a single exponential function. (n = 5–7 cells).

3.4. High Frequency Stimulation

For optogenetics application, reliable neuronal activation would require large and stable activity of the light-gated ion channel. In addition, rapid channel-closing is desired for optical stimulation at a high frequency. Thus, we compared the photocurrent from *Gt*_CCR4- and *Cr*_ChR2-expressing cells with three different light frequencies. As shown in Figure 4A, activation of *Gt*_CCR4 at 10 Hz light at 9.67 mW/mm^2 generated high temporal peak currents which almost fully decayed before the next illumination. Peak amplitude remained unchanged because of a small inactivation, as shown in Figure 1A. On the other hand, peak current amplitude immediately decayed less after the initial stimulation in *Cr*_ChR2-expressing cells (Figure 4B). Stimulation of *Gt*_CCR4 at 20 and 50 Hz still retained a high level of peak amplitudes, although each peak did not decay completely (Figure 4C,E). Current inactivation was observed in *Cr*_ChR2-expressing cells at each frequency (20 and 50 Hz)

(Figure 4D,F). Figure 4G summarizes the residual current level before the next stimulation at each light frequency. At 10 Hz, both *Gt*_CCR4 and *Cr*_ChR2 showed a very low level of the current level, indicating that channels were almost shut off. As frequency increased to 20 Hz and 50 Hz, significant residual currents were observed which exceeded 50% in *Gt*_CCR4 at 50 Hz whereas *Cr*_ChR2 showed a slightly lower residual current, probably because of faster channel kinetics.

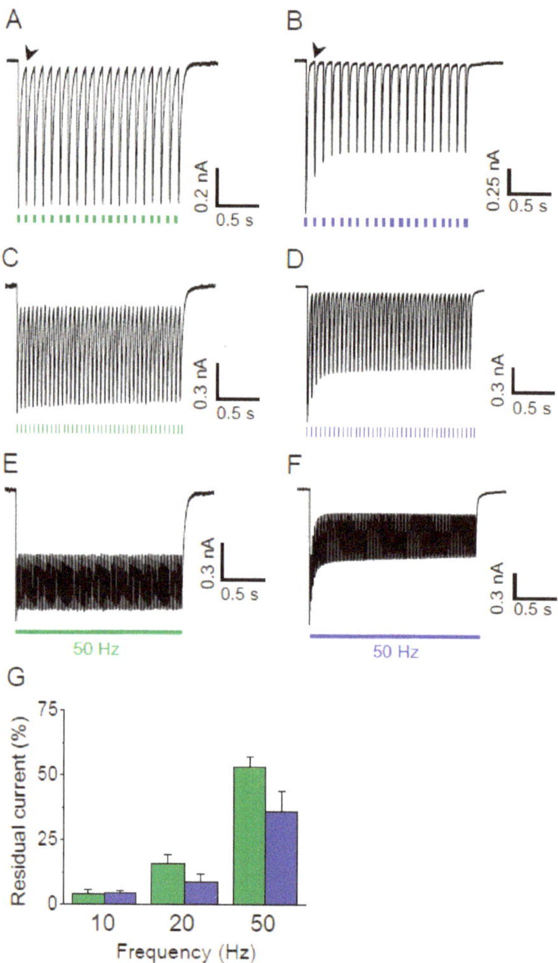

Figure 4. High frequency stimulation of *Gt*_CCR4 and *Cr*_ChR2. (**A**,**C**,**E**) *Gt*_CCR4 in ND7/23 cells was stimulated by green LED light with a frequency of 10, 20 or 50 Hz as indicated by colored dots or a line under each trace. Membrane voltage was clamped at −60 mV. (**B**,**D**,**F**) Similarly, *Cr*_ChR2 was stimulated by blue LED light with a frequency of 10, 20 or 50 Hz as indicated by colored dots or a line under each trace. (**G**) Residual tail current at three different light frequencies is summarized. Residual tail current is the current amplitude right before the next photo stimulation indicated by an arrowhead in **A**,**B**. The current value was normalized to the peak amplitude as 100%. A standard solution was used. (n = 3 cells).

3.5. Gt_CCR4 Is Inactivated by 590 nm Light But Fully Reactivated by Blue Light

As already shown in Figure 1A, one of the obvious characteristics of *Gt*_CCR4 is the small inactivation upon 530 nm illumination which is close to λ_{max} (525 nm). Here we compared the current shape by illumination of 530 nm and 590 nm light (Figure 5A–C). Repetitive illuminations of 530 nm light three times gave almost the same current shape, which has a high steady state level (Figure 5A). Upon illumination of 590 nm light, the current slowly decayed into a small steady state level, which was further reduced after a second illumination of 590 nm LED following several seconds of a dark period (Figure 5B,D). Even after 30 sec or 2 min, the inactivated current did not recover (Figure S1A,B). These observations suggest that *Gt*_CCR4 possesses a long-lived inactivated state which could accumulate in 590 nm light. Only illuminating with 530 nm allowed for a full recovery to the original steady state level (Figure 5C,E).

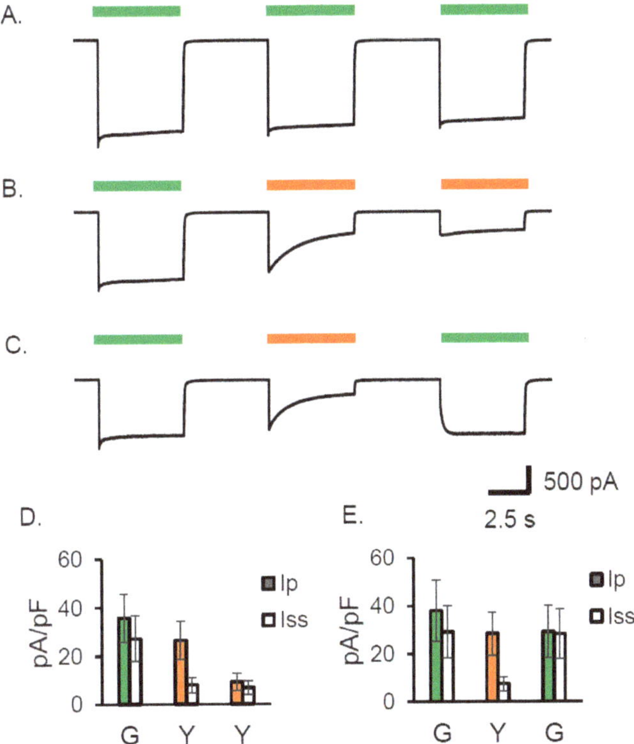

Figure 5. *Gt*_CCR4 is inactivated by light with a longer wavelength. Standard solutions were used. Membrane voltage was clamped at −60 mV. Photocurrent by 530 nm light (7.44 mW/mm^2) in (**A**) reached a steady state level after transient peak current. Repetitive stimulations gave almost identical current shapes. (**B**) Slow inactivation of the photocurrent was observed by illumination with 590 nm light (7.44 mW/mm^2). The current was further reduced when illuminated twice with 590 nm light. (**C**) After 590 nm inactivation, 530 nm light fully reactivated the photocurrent to the original steady state level. (**D**) Photocurrent density from the measurement shown in (**B**). The photocurrent density upon exposure to the first 530 nm light (shown in G), and the second and third 590 nm light (Y), is shown. Ip; transient peak component. Iss; steady state component. (**E**) Photocurrent density from the measurement shown in C. The photocurrent density upon exposure to the first 530 nm light (shown in G), the second 590 nm light (Y), and the third 530 nm light (shown in G), is shown. (n = 4 cells).

4. Discussion

In this study, we aimed to elucidate the ion-channel properties of a recently discovered light-gated cation channel *Gt*_CCR4 from a cryptophyte and compare it to well-known *Cr*_ChR2 from *Chlamydomonas reinhardtii* by using an electrophysiological method. In ND7/23 cells, *Gt*_CCR4 showed a large current density (Figures 1C and 2A,B). Inactivation of the photocurrent obtained from *Gt*_CCR4 was smaller than in *Cr*_ChR2. In other words, a large current was observed under constant light (Figure 1A,B). This was also obvious in the current trace after illumination at a high frequency (Figure 4). These characteristics promise stable and reproducible stimulation of neuronal excitability by *Gt*_CCR4.

The light sensitivity of *Gt*_CCR4 is higher than that of *Cr*_ChR2 with a particular steady state component (I_{ss}) (Figure 1E,F). ChR variants with high light sensitivity have already been developed [19,20], but those have a long channel life time with at least two orders of magnitude or even much longer. Therefore, these are inappropriate for high-frequency light stimulation. In contrast, *Gt*_CCR4 has a short open life time of 25–30 ms, which is about the same range as *Cr*_ChR2, i.e., 10–15 ms (Figure 1D). Together, *Gt*_CCR4 is light sensitive and useful as an optogenetics tool with high time resolution. Optical irradiation causes heat and elevates temperature by 0.2~2 °C, especially in cranial nerve experiments [37]. Moreover, it has been demonstrated that the rise in temperature suppressed neuronal spiking in multiple brain regions, serving as a warning of the use of strong light for neuronal stimulation. Such an undesirable artefact has to be avoided by lowering light intensity, while effective depolarization has to be stably maintained. *Gt*_CCR4 has the potential for overcoming this problem.

H^+ permeability is high for *Cr*_ChR2 [8]. Permeability for Ca^{2+} has been reported, and not only for monovalent cations such as Na^+ and K^+ [8,21]. On the other hand, *Gt*_CCR4 showed high selectivity in monovalent metal cations and low H^+ permeability. The permeability of a divalent cation such as Ca^{2+} seems to be very low or negligible. The position that is important to ion selectivity has been studied in *Cr*_ChR2. E90 in the central gate is crucial for cation/anion selection [23]. L132 in TM3 influences on Ca^{2+} permeability [21]. Duan and coworkers recently demonstrated that D156H and D156C mutation increase permeability for Na^+ and K^+ [38]. The outer gate in Chrimson (E139) on the extracellular side is important for Na^+ extrusion [22]. These key residues are not conserved in *Gt*_CCR4, implying that a different ion selection property resides in DTD channels. It would be necessary to study selectivity based on variant analysis and structural information in the future. Considering its application in optogenetics, *Gt*_CCR4 would not cause a significant change to pH in the cell membrane because of its very low H^+ permeability, which could be advantageous when an unknown effect by pH needs to be prevented. To enable optical stimulation without improper calcium signaling, *Gt*_CCR4 might work better than *Cr*_ChR2.

A single turnover photocurrent of *Gt*_CCR4 by laser irradiation provided a time constant (τ_{off}) of 15–20 ms, which is smaller than that obtained by constant light (25–30 ms). This suggests two processes for channel opening and shutting. A dual photocycle model was indeed proposed for *Cr*_ChR2 [39]. It is expected that a similar reaction is caused in *Gt*_CCR4, but more experiments are needed to prove this.

We found a characteristic inactivation of channel activity by long wavelength absorption in *Gt*_CCR4 (Figure 5). Since inactivation lasted at least a few minutes, formation of a stable intermediate with long wavelength absorption is anticipated. Alternatively, *Gt*_CCR4 exhibits a photochromic property that is seen in the photocycle of *Anabaena* sensory rhodopsin [40]. Such photochromism or desensitization was also observed in chlorophyte channelrhodopsins [41,42]. We are now focusing on understanding the reaction mechanism in greater depth via a spectroscopic experiment. In conclusion, we here elucidated the cation channel properties of *Gt*_CCR4. Its high conductance and cation selectivity without significant inactivation would be an appropriate set of features for optogenetics applications. We are currently assessing the feasibility of *Gt*_CCR4 as an optical stimulator in cultured neurons.

Supplementary Materials: The following are available online at http://www.mdpi.com/2076-3417/9/17/3440/s1, Table S1 Amino acid alignments of bacteriorhodopsin (BR), *Cr*_ChR2 and *Gt*_CCR4. Figure S1: Ca^{2+} photocurrents of *Gt*_CCR4 and *Cr*_ChR2. Figure S2 Gt_CCR4 has a long-lived and long wavelength-absorbing inactivated state.

Author Contributions: Conceptualization, All authors; Data Curation, S.S. and S.H.; Validation, S.S., S.H. and S.P.T.; Writing—Original Draft Preparation, S.P.T.; Writing—Review & Editing, S.P.T.; Supervision, H.K. and S.P.T.

Funding: This work was funded by the Japanese Ministry of Education, Culture, Sports, Science and Technology (25104009, 15H02391 to H.K and 18K06109 to S.P.T.), a JST CREST grant (JPMJCR1753 to H.K.), and a JST PRESTO grant (JPMJPR1688 to S.P.T). S.H. is a Research Fellow of the Japan Society for the Promotion of Science (JSPS Research Fellow).

Acknowledgments: We thank Ryoko Nakamura for her excellent technical support.

Conflicts of Interest: The authors declare no conflict of interest.

References

1. Ernst, O.P.; Lodowski, D.T.; Elstner, M.; Hegemann, P.; Brown, L.S.; Kandori, H. Microbial and animal rhodopsins: Structures, functions, and molecular mechanisms. *Chem. Rev.* **2014**, *114*, 126–163. [CrossRef] [PubMed]
2. Oesterhelt, D.; Stoeckenius, W. Rhodopsin-like protein from the purple membrane of Halobacterium halobium. *Nat. New Biol.* **1971**, *233*, 149–152. [CrossRef] [PubMed]
3. Matsuno-Yagi, A.; Mukohata, Y. Two Possible Roles of Bacteriorhodopsin; a Comparative Study of Strains of Halobacz'eril'm Halobium Differing in Pigmentation. *Biochem. Biophys. Res. Commun.* **1977**, *78*, 237–243. [CrossRef]
4. Inoue, K.; Ono, H.; Abe-Yoshizumi, R.; Yoshizawa, S.; Ito, H.; Kogure, K.; Kandori, K. A light-driven sodium ion pump in marine bacteria. *Nat. Commun.* **2013**, *4*, 1678. [CrossRef] [PubMed]
5. Inoue, K.; Ito, S.; Kato, Y.; Nomura, Y.; Shibata, M.; Uchihashi, T.; Tsunoda, S.P.; Kandori, H. A natural light-driven inward proton pump. *Nat. Commun.* **2016**, *7*, 13415. [CrossRef] [PubMed]
6. Polovinkin, V.; Rodriguez-Valera, F.; Bueldt, G.; Kovalev, K.; Alekseev, A.; Gordeliy, V.; Chizhov, I.; Juettner, J.; Bamann, C.; Manstein, D.J.; et al. Inward H + pump xenorhodopsin: Mechanism and alternative optogenetic approach. *Sci. Adv.* **2017**, *3*, e1603187.
7. Nagel, G.; Ollig, D.; Fuhrmann, M.; Kateriya, S.; Musti, A.M.; Bamberg, E.; Hegemann, P. Channelrhodopsin-1: A Light-Gated Proton Channel in Green Algae. *Science* **2002**, *296*, 2395–2398. [CrossRef]
8. Nagel, G.; Szellas, T.; Huhn, W.; Kateriya, S.; Adeishvili, N.; Berthold, P.; Ollig, D.; Hegemann, P.; Bamberg, E. Channelrhodopsin-2, a directly light-gated cation-selective membrane channel. *Proc. Natl. Acad. Sci. USA* **2003**, *100*, 13940–13945. [CrossRef]
9. Kato, H.E.; Zhang, F.; Yizhar, O.; Ramakrishnan, C.; Nishizawa, T.; Hirata, K.; Ito, J.; Aita, Y.; Tsukazaki, T.; Hayashi, S.; et al. Crystal structure of the channelrhodopsin light-gated cation channel. *Nature* **2012**, *482*, 369–374. [CrossRef]
10. Volkov, O.; Kovalev, K.; Polovinkin, V.; Borshchevskiy, V.; Bamann, C.; Astashkin, R.; Marin, E.; Popov, A.; Balandin, T.; Willbold, D.; et al. Structural insights into ion conduction by channelrhodopsin 2. *Science* **2017**, *358*, pii: eaan8862. [CrossRef]
11. Boyden, E.S.; Zhang, F.; Bamberg, E.; Nagel, G.; Deisseroth, K. Millisecond-timescale, genetically targeted optical control of neural activity. *Nat. Neurosci.* **2005**, *8*, 1263–1268. [CrossRef]
12. Ishizuka, T.; Kakuda, M.; Araki, R.; Yawo, H. Kinetic evaluation of photosensitivity in genetically engineered neurons expressing green algae light-gated channels. *Neurosci. Res.* **2006**, *54*, 85–94. [CrossRef]
13. Schneider, F.; Grimm, C.; Hegemann, P. Biophysics of Channelrhodopsin. *Annu. Rev. Biophys.* **2015**, *44*, 167–186. [CrossRef]
14. Prigge, M.; Schneider, F.; Tsunoda, S.P.; Shilyansky, C.; Wietek, J.; Deisseroth, K.; Hegemann, P. Color-tuned channelrhodopsins for multiwavelength optogenetics. *J. Biol. Chem.* **2012**, *287*, 31804–31812. [CrossRef]
15. Wang, H.; Sugiyama, Y.; Hikima, T.; Sugano, E.; Tomita, H.; Takahashi, T.; Ishizuka, T.; Yawo, H. Molecular determinants differentiating photocurrent properties of two channelrhodopsins from Chlamydomonas. *J. Biol. Chem.* **2009**, *284*, 5685–5696. [CrossRef]
16. Tsunoda, S.P.; Hegemann, P. Glu 87 of channelrhodopsin-1 causes pH-dependent color tuning and fast photocurrent inactivation. *Photochem. Photobiol.* **2009**, *85*, 564–569. [CrossRef]

17. Klapoetke, N.C.; Murata, Y.; Kim, S.S.; Pulver, S.R.; Birdsey-Benson, A.; Cho, Y.K.; Morimoto, T.K.; Chuong, A.S.; Carpenter, E.J.; Tian, Z.; et al. Independent optical excitation of distinct neural populations. *Nat. Methods* **2014**, *11*, 338–346. [CrossRef]
18. Govorunova, E.G.; Sineshchekov, O.A.; Li, H.; Janz, R.; Spudich, J.L. Characterization of a highly efficient blue-shifted channelrhodopsin from the marine alga platymonas subcordiformis. *J. Biol. Chem.* **2013**, *288*, 29911–29922. [CrossRef]
19. Berndt, A.; Yizhar, O.; Gunaydin, L.A.; Hegemann, P.; Deisseroth, K. Bi-stable neural state switches. *Nat. Neurosci.* **2009**, *12*, 229–234. [CrossRef]
20. Yizhar, O.; Fenno, L.E.; Prigge, M.; Schneider, F.; Davidson, T.J.; Ogshea, D.J.; Sohal, V.S.; Goshen, I.; Finkelstein, J.; Paz, J.T.; et al. Neocortical excitation/inhibition balance in information processing and social dysfunction. *Nature* **2011**, *477*, 171–178. [CrossRef]
21. Kleinlogel, S.; Feldbauer, K.; Dempski, R.E.; Fotis, H.; Wood, P.G.; Bamann, C.; Bamberg, E. Ultra light-sensitive and fast neuronal activation with the Ca^{2+}-permeable channelrhodopsin CatCh. *Nat. Neurosci.* **2011**, *14*, 513–518. [CrossRef]
22. Vierock, J.; Grimm, C.; Nitzan, N.; Hegemann, P. Molecular determinants of proton selectivity and gating in the red-light activated channelrhodopsin Chrimson. *Sci. Rep.* **2017**, *7*, 1–15. [CrossRef]
23. Wietek, J.; Wiegert, J.S.; Adeishvili, N.; Schneider, F.; Watanabe, H.; Tsunoda, S.P.; Vogt, A.; Elstner, M.; Oertner, T.G.; Hegemann, P. Conversion of channelrhodopsin into a light-gated chloride channel. *Science* **2014**, *344*, 409–412. [CrossRef]
24. Berndt, A.; Lee, S.Y.; Ramakrishnan, C.; Deisseroth, K. Structure-guided transformation of channelrhodopsin into a light-activated chloride channel. *Science* **2014**, *344*, 420–424. [CrossRef]
25. Govorunova, E.G.; Sineshchekov, O.A.; Janz, R.; Liu, X.; Spudich, J.L. NEUROSCIENCE. Natural light-gated anion channels: A family of microbial rhodopsins for advanced optogenetics. *Science* **2015**, *349*, 647–650. [CrossRef]
26. Kim, Y.S.; Kato, H.E.; Yamashita, K.; Ito, S.; Inoue, K.; Ramakrishnan, C.; Fenno, L.E.; Evans, K.E.; Paggi, J.M.; Dror, R.O.; et al. Crystal structure of the natural anion-conducting channelrhodopsin GtACR1. *Nature* **2018**, *561*, 343–348. [CrossRef]
27. Kato, H.E.; Kim, Y.S.; Paggi, J.M.; Evans, K.E.; Allen, W.E.; Richardson, C.; Inoue, K.; Ito, S.; Ramakrishnan, C.; Fenno, L.E.; et al. Structural mechanisms of selectivity and gating in anion channelrhodopsins. *Nature* **2018**, *561*, 349–371. [CrossRef]
28. Spudich, J.L.; Li, H.; Huang, C.-Y.; Wang, M.; Schafer, C.T.; Sineshchekov, O.A.; Zheng, L.; Govorunova, E.G. Crystal structure of a natural light-gated anion channelrhodopsin. *Elife* **2019**, *8*, 1–21.
29. Govorunova, E.G.; Sineshchekov, O.A.; Spudich, J.L. Structurally Distinct Cation Channelrhodopsins from Cryptophyte Algae. *Biophys. J.* **2016**, *110*, 2302–2304. [CrossRef]
30. Yamauchi, Y.; Konno, M.; Ito, S.; Tsunoda, S.P.; Inoue, K.; Kandori, H. Molecular properties of a DTD channelrhodopsin from Guillardia theta. *Biophys. Physicobiol.* **2017**, *14*, 57–66. [CrossRef]
31. Sugiyama, Y.; Wang, H.; Hikima, T.; Sato, M.; Kuroda, J.; Takahashi, T.; Ishizuka, T.; Yawo, H. Photocurrent attenuation by a single polar-to-nonpolar point mutation of channelrhodopsin-2. *Photochem. Photobiol. Sci.* **2009**, *8*, 328–336. [CrossRef]
32. Dawydow, A.; Gueta, R.; Ljaschenko, D.; Ullrich, S.; Hermann, M.; Ehmann, N.; Gao, S.; Fiala, A.; Langenhan, T.; Nagel, G.; et al. Channelrhodopsin-2-XXL, a powerful optogenetic tool for low-light applications. *Proc. Natl. Acad. Sci. USA* **2014**, *111*, 13972–13977. [CrossRef]
33. Nack, M.; Radu, I.; Gossing, M.; Bamann, C.; Bamberg, E.; von Mollard, G.F.; Heberle, J.; Hofkens, J.; Pozzo, J.L.; Tosic, O.; et al. The DC gate in Channelrhodopsin-2: Crucial hydrogen bonding interaction between C128 and D156. *Photochem. Photobiol. Sci.USA* **2010**, *9*, 194–198. [CrossRef]
34. Sineshchekov, O.A.; Govorunova, E.G.; Li, H.; Spudich, J.L. Bacteriorhodopsin-like channelrhodopsins: Alternative mechanism for control of cation conductance. *Proc. Natl. Acad. Sci.* **2017**, *114*, E9512–E9519. [CrossRef]
35. Lorenz-Fonfria, V.A.; Resler, T.; Krause, N.; Nack, M.; Gossing, M.; Fischer von Mollard, G.; Bamann, C.; Bamberg, E.; Schlesinger, R.; Heberle, J. Transient protonation changes in channelrhodopsin-2 and their relevance to channel gating. *Proc. Natl. Acad. Sci.* **2013**, *110*, E1273–E1281. [CrossRef]

36. Wood, J.N.; Bevan, S.J.; Coote, P.R.; Dunn, P.M.; Harmar, A.; Hogan, P.; Latchman, D.S.; Morrison, C.; Rougon, G.; Theveniau, M.; et al. Novel Cell Lines Display Properties of Nociceptive Sensory Neurons. *Proc. R. Soc. B Biol. Sci.* **1990**, *241*, 187–194.
37. Owen, S.F.; Liu, M.H.; Kreitzer, A.C. Thermal constraints on in vivo optogenetic manipulations. *Nat. Neurosci.* **2019**, *22*, 1061–1065. [CrossRef]
38. Duan, X.; Nagel, G.; Gao, S. Mutated Channelrhodopsins with Increased Sodium and Calcium Permeability. *Appl. Sci.* **2019**, *9*, 664. [CrossRef]
39. Kuhne, J.; Vierock, J.; Tennigkeit, S.A.; Dreier, M.-A.; Wietek, J.; Petersen, D.; Gavriljuk, K.; El-Mashtoly, S.F.; Hegemann, P.; Gerwert, K. Unifying photocycle model for light adaptation and temporal evolution of cation conductance in channelrhodopsin-2. *Proc. Natl. Acad. Sci USA* **2019**, *116*, 9380–9389. [CrossRef]
40. Kawanabe, A.; Furutani, Y.; Jung, K.H.; Kandori, H. Photochromism of Anabaena sensory rhodopsin. *J. Am. Chem. Soc.* **2007**, *129*, 8644–8649. [CrossRef]
41. Zamani, A.; Sakuragi, S.; Ishizuka, T.; Yawo, H. Kinetic characteristics of chimeric channelrhodopsins implicate the molecular identity involved in desensitization. *Biophys. Physicobiol.* **2017**, *14*, 13–22. [CrossRef]
42. Lin, J.Y. A User's Guide to Channelrhodopsin Variants: Features, Limitations and Future Developments. *Exp. Physiol.* **2011**, *96*, 19–25. [CrossRef]

Article

Mutated Channelrhodopsins with Increased Sodium and Calcium Permeability

Xiaodong Duan, Georg Nagel * and Shiqiang Gao *

Botanik I, Julius-Maximilians-Universität Würzburg, Biozentrum, Julius-von-Sachs-Platz 2, D-97082 Würzburg, Germany; xiaodong.duan@stud-mail.uni-wuerzburg.de

* Correspondence: nagel@uni-wuerzburg.de (G.N.); gao.shiqiang@uni-wuerzburg.de (S.G.)

Received: 14 January 2019; Accepted: 12 February 2019; Published: 15 February 2019

Featured Application: This study provides optogenetic tools with superior photocurrent amplitudes and high Na^+ and Ca^{2+} conductance.

Abstract: (1) Background: After the discovery and application of *Chlamydomonas reinhardtii* channelrhodopsins, the optogenetic toolbox has been greatly expanded with engineered and newly discovered natural channelrhodopsins. However, channelrhodopsins of higher Ca^{2+} conductance or more specific ion permeability are in demand. (2) Methods: In this study, we mutated the conserved aspartate of the transmembrane helix 4 (TM4) within Chronos and *Ps*ChR and compared them with published ChR2 aspartate mutants. (3) Results: We found that the ChR2 D156H mutant (XXM) showed enhanced Na^+ and Ca^{2+} conductance, which was not noticed before, while the D156C mutation (XXL) influenced the Na^+ and Ca^{2+} conductance only slightly. The aspartate to histidine and cysteine mutations of Chronos and *Ps*ChR also influenced their photocurrent, ion permeability, kinetics, and light sensitivity. Most interestingly, *Ps*ChR D139H showed a much-improved photocurrent, compared to wild type, and even higher Na^+ selectivity to H^+ than XXM. *Ps*ChR D139H also showed a strongly enhanced Ca^{2+} conductance, more than two-fold that of the CatCh. (4) Conclusions: We found that mutating the aspartate of the TM4 influences the ion selectivity of channelrhodopsins. With the large photocurrent and enhanced Na^+ selectivity and Ca^{2+} conductance, XXM and *Ps*ChR D139H are promising powerful optogenetic tools, especially for Ca^{2+} manipulation.

Keywords: optogenetics; channelrhodopsins; sodium; calcium; DC gate

1. Introduction

Channelrhodopsins were first discovered and characterized from *C. reinhardtii* [1,2]. After the showing of light-switched large passive cation conductance in HEK293 and BHK cells by Nagel et al., the ChR2 (*C. reinhardtii* channelrhodopsin-2) was immediately applied in neuroscience by several independent groups for studies in hippocampal neurons [3,4], *Caenorhabditis elegans* [5], inner retinal neurons [6], and PC12 cells [7]. H134R (histidine to arginine mutation at position 134) was the first ChR2 gain-of-function mutant which showed enhanced plasma membrane expression and larger stationary photocurrents in comparison to ChR2 wild type [5].

Other variants came out in rapid sequence, either of natural origin or mutated and engineered. The calcium translocating channelrhodopsin CatCh (a ChR2 leucine to cysteine mutation at position 132, L132C) showed improved Ca^{2+} conductivity together with a larger photocurrent and higher light sensitivity [8]. Newly discovered Chronos (*Stigeoclonium helveticum* channelrhodopsin = *Sh*ChR) and Chrimson (*Cn*ChR1 from *Chlamydomonas noctigama*) showed a faster channel closing and a red-shifted action spectrum, respectively [9]. An E90R (glutamate to arginine mutation at position 90) point mutation could extend the cation conductance of ChR2 to additional anion conductance [10]. Naturally, very specific anion conductive channelrhodopsins, *Gt*ACR1 and *Gt*ACR2 (*Guillardia theta* anion channelrhodopsin 1 and 2), were discovered afterwards [11].

Mutation of ChR2 C128 (cysteine at position 128) to threonine (T), alanine (A), and serine (S) slowed the closing kinetics dramatically [12–14]. Mutation of ChR2 D156 (cysteine at position 156) to alanine also decreased the closing kinetics [14]. Spectral studies suggested that a putative hydrogen bond between C128 and D156 could be an important structural determinant of the channel's closing reaction [14] or might represent the valve of the channel [15]. Thus, hydrogen bond-linked D156 and C128 was proposed as the putative gate buried in the membrane ("DC gate") [14,15]. But the first channelrhodopsin structure was a chimaera (C1C2) of truncated ChR1 and ChR2, and the distances between C167 (corresponding to C128 in ChR2) and D195 (corresponding to D156 in ChR2) are too far away to be associated by a hydrogen bond [16]. However, the recently solved structure of wild type ChR2 revealed a water molecule between C128 and D156 to bridge them, indeed, by hydrogen bonds [17].

Mutation of the DC gate has a strong effect on the open channel lifetime. The ChR2 D156C (aspartate to cysteine mutation at position 134) mutant (XXL) generated very large photocurrents and is 1000-fold more light-sensitive than wild type ChR2 in *Drosophila* larvae [18]. The ChR2 D156H (aspartate to histidine mutation at position 134) mutant (XXM) also showed a superior photo stimulation efficiency with faster kinetics than XXL, which made it an ideal optogenetic tool for *Drosophila* neurobiological studies [19].

The aspartate D156 in ChR2 is located close to the protonated retinal Schiff base (RSBH+). Thus, mutations of D156 logically have strong effects on the open channel lifetime by influencing the protonation state of the retinal Schiff base. However, the water-bridged C128 and D156 are not in the putative ion pore proposed by Volkov et al. [17]. And no attention had been paid to the potential changes of ion selectivity by DC gate mutations.

In this study, we compared the ion selectivity of our previously published XXL and XXM and found that XXM showed a four-fold increased Na^+ selectivity over H^+ together with a two-fold increased K^+ selectivity over H^+, compared to wild type ChR2. Based on this finding, we made further aspartate to histidine and cysteine mutations of *Ps*ChR (*Platymonas subcordiformis* channelrhodopsin) [20] and Chronos [9]. *Ps*ChR wild type was already reported to be highly Na^+ conductive and indeed showed a six-fold increased Na^+ and K^+ selectivity over H^+ compared to wild type ChR2 in our measurements. But the D139H mutation of *Ps*ChR further increased the Na^+ selectivity over H^+ five-fold. Furthermore, *Ps*ChR D139H showed a 5-fold larger photocurrent than *Ps*ChR wt.

We further compared the Ca^{2+} permeability of these mutants. XXM showed an increased Ca^{2+} current compared to CatCh [8]. *Ps*ChR wild type already showed a good Ca^{2+} current, but the D139H mutation further increased the Ca^{2+} current. We concluded that the mutant *Ps*ChR D139H would be a powerful tool for optogenetic Ca^{2+} manipulation.

2. Materials and Methods

2.1. Plasmids and RNA Generation for Xenopus Laevis Oocyte Expression

ChR2, XXM, and XXL in the pGEMHE vector were described in previous studies [18,19,21]. PsChR (from Platymonas subcordiformis, Accession No.: JX983143) and Chronos (from Stigeoclonium helveticum, Accession No.: KF992040) were synthesized by GeneArt Strings DNA Fragments (Life Technologies, Thermo Fisher Scientific), according to the published amino acid sequences, with the codon usage optimized to Mus musculus. The synthesized DNA segment was inserted into the pGEMHE vector with N-terminal BamHI and C-terminal XhoI restriction sites. Yellow fluorescent protein (YFP), together with a plasma membrane trafficking signal (KSRITSEGEYIPLDQIDINV) [22] beforehand and an ER export signal (FCYENEV) [22] afterward, the YFP was attached to the C-terminal end. Mutations were made by QuikChange Site-Directed Mutagenesis. The sequence was confirmed by DNA sequencing. Plasmids were linearized by NheI digestion and used for in vitro generation of cRNA with the AmpliCap-MaxT7 High Yield Message Maker Kit (Epicentre Biotechnologies).

2.2. *Two-Electrode Voltage-Clamp Recordings of Xenopus Laevis Oocytes*

cRNA-injected oocytes were incubated in ND96 solution (96 mM NaCl, 5 mM KCl, 1 mM $MgCl_2$, 1 mM $CaCl_2$, 5 mM HEPES, pH 7.4) containing 1 μM all-trans-retinal at 16 °C. Two-electrode voltage-clamp (TEVC) recordings were performed with solutions, as indicated in figures, at room temperature. For experiments with external Ca^{2+}, we blocked activation of the Ca^{2+}-activated endogenous chloride channels of oocytes by 1,2-bis(o-aminophenoxy)ethane-N,N,N′,N′-tetraacetic acid (BAPTA) injection. We injected 50 nl 200 mM of the fast Ca^{2+} chelator BAPTA (potassium–salt) into each oocyte (~10 mM final concentration in the oocyte), incubated for 90 mins at 16 °C and then performed the TEVC measurement at room temperature. Twenty nanograms of cRNA were injected into *Xenopus* oocyte for all the constructs. Photocurrents were measured two days after injection. For Figures 1 and 2, measurements were performed in standard solution with $BaCl_2$ instead of $CaCl_2$ (110 mM NaCl, 5 mM KCl, 2 mM $BaCl_2$, 1 mM $MgCl_2$, 5 mM HEPES and pH 7.6).

2.3. *Light Stimulation*

Illumination conditions were different, considering the published action spectra of ChR2, Chronos, and *Ps*ChR: 473 nm for ChR2, XXM, and XXL; 445 nm for *Ps*ChR, *Ps*ChR D139H, and *Ps*ChR D139C; 532 nm for Chronos, Chronos D173H, and Chronos D173C. Lasers were from Changchun New Industries Optoelectronics Technology. Light power was set to 5 mW/mm^2, except for the light sensitivity measurement. The light intensities were measured with a PLUS 2 Power & Energy Meter (LaserPoint s.r.l). For light sensitivity measurement, applied light intensities ranged from 1.7 to 5000 μW/mm^2.

2.4. *Protein Quantification by Fluorescence*

All expression levels of channelrhodopsin variants in oocytes were quantified by the fluorescence emission values of the YFP-tagged protein. Fluorescence emission was measured at 538 nm by a Fluoroskan Ascent microplate fluorometer (Thermo Scientific) with 485 nm excitation.

2.5. *Fluorescence Imaging*

Fluorescence pictures of *Xenopus* oocytes were taken under 5x objective with a Leica DM6000 confocal microscope after two days' expression. Oocytes were put in a 35 x 10 mm petri dish (Greiner GBO) containing ND96 for imaging. Excitation was done using 496 nm laser light. Fluorescence emission was detected from 520 nm to 585 nm.

2.6. *Data Processing*

pClamp 7.0 was used to read out the photocurrent. Figures 1a, 2, 3, 4, 5a and A1a were made with OriginPro 2017. Figures 1b, 5b and A1b were made with GraphPad Prism. Tables 1 and 2 were made with Microsoft Excel. Sequence alignment in Appendix B was performed by BioEdit. Closing time was determined by biexponential fit. Light sensitivity curves were fitted with Hill equation. All values were plotted or presented with mean values, and error bars represent the standard deviations (SD) or standard error mean (SEM), as indicated in each figure. Statistical analysis was done by *t* test within GraphPad Prism. Differences were considered significant at $p < 0.05$. *** = $p < 0.001$, ** = $p < 0.01$, * = $p < 0.05$.

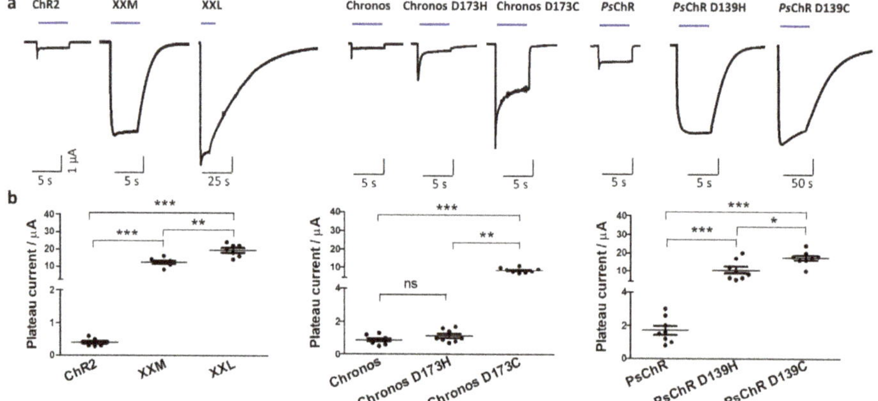

Figure 1. Comparison of ChR2, *Ps*ChR, and Chronos variants. (**a**) Representative photocurrent traces of ChR2, *Ps*ChR, Chronos, and their mutants. (**b**) Comparison of stationary photocurrents of above channelrhodopsin variants. All data points were plotted in the figure and mean ± standard error mean (SEM) (n = 7–8) was indicated.

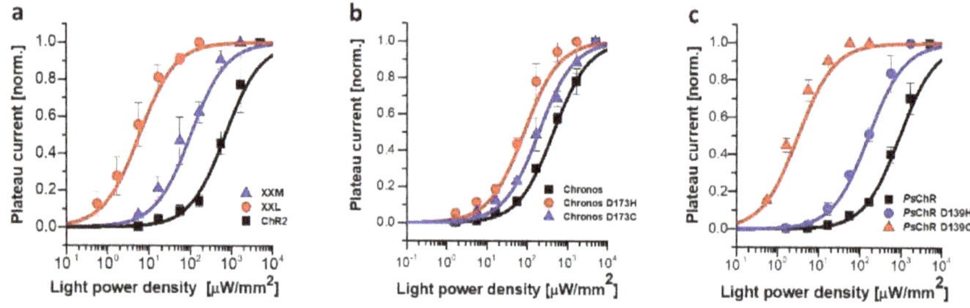

Figure 2. Comparing the light sensitivities of ChR2, *Ps*ChR, and Chronos variants. (**a**) 5 s, 30 s, and 100 s continuous 473 nm blue light illumination were applied to ChR2, XXM and XXL, respectively. Photocurrent at maximum light power density of each oocyte was normalized as 1. (**b**) 5 s continuous 532 nm blue light illumination were applied to Chronos, Chronos D173H, and Chronos D173C. Photocurrent at maximum light power density of each oocyte was normalized as 1. (**c**) 5 s, 30 s, and 100 s continuous 445 nm blue light illumination were applied to *Ps*ChR, *Ps*ChR D139H, and *Ps*ChR D139C. Photocurrent at maximum light power density of each oocyte was normalized as 1. Photocurrents of XXM, XXL, *Ps*ChR D139H, and *Ps*ChR D139C were measured at −60 mV because of the larger current and slower kinetics; other constructs were measured at −100 mV. Data points were presented as mean ± SD, n = 3–4.

3. Results

3.1. Mutating the Conserved Aspartate of TM Helix 4 Influences the Expression, Photocurrent, and Kinetics

Similar to previously published results [18,19], fluorescence measurements of whole oocyte membranes with YFP-tagged XXM and XXL showed a ~ three-fold increased expression level, compared to ChR2 (Figure A1). The steady-state photocurrents were increased ~30- and ~48-fold for XXM (D156H) and XXL (D156C), respectively, compared to ChR2 (Figure 1 and Table 1). The enhanced photocurrent might have been a comprehensive outcome of the higher plasma membrane expression level, higher light-sensitivity, or increased single channel conductance. Both XXM and XXL showed much-prolonged closing kinetics, leading to higher light-sensitivity (Figure 1a and Table 1).

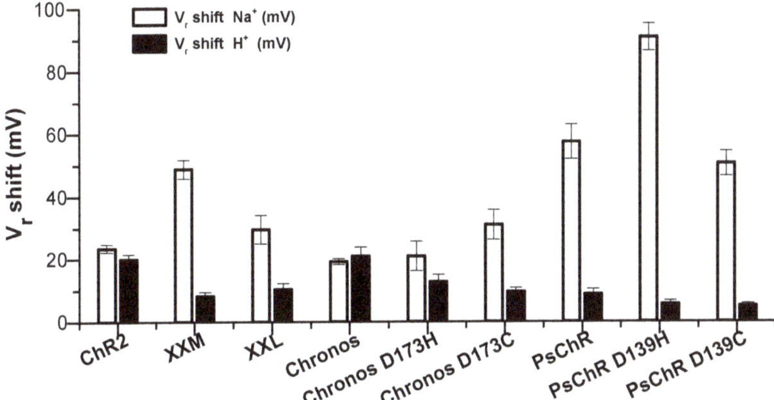

Figure 3. Comparison of Na^+ and H^+ permeabilities of ChR2, Chronos, *Ps*ChR, and their mutants. Reversal potentials (V_r) were determined after photocurrent measurements from −90 mV to + 10 mV. Reversal potential shift for Na^+ was calculated by the reversal potential differences in two outside buffers containing 120 mM NaCl pH 7.6 and 1 mM NaCl pH 7.6. Reversal potential shift for H^+ was determined by the reversal potential differences in two outside buffers of pH 7.6 and 9.6 containing 120 mM NaCl. Data points were presented as mean ± SD, *n* = 4–6.

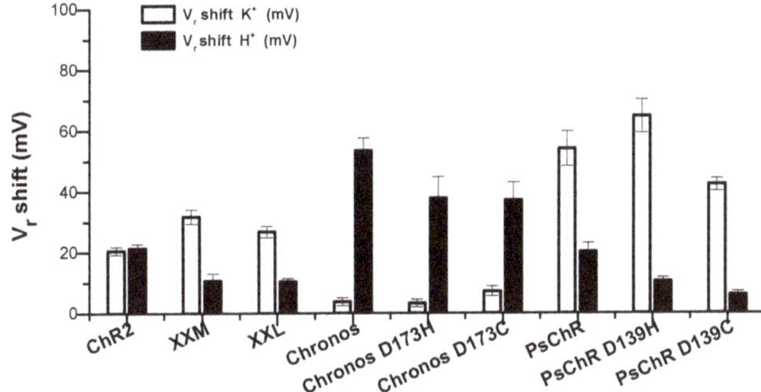

Figure 4. Comparison of K^+ and H^+ permeabilities of ChR2, Chronos, *Ps*ChR, and their mutants. Reversal potentials were determined after photocurrent measurements from −90 mV to +10 mV. Reversal potential shift for K^+ was calculated by the reversal potential differences in two outside buffers containing 120 mM KCl pH 7.6 and 1 mM KCl pH 7.6. Reversal potential shift for H^+ was determined by the reversal potential differences in two outside buffers of pH 7.6 and 9.6 containing 120 mM KCl. Data points were presented as mean ± SD, *n* = 4–6.

We further synthesized Chronos [9] and *Ps*ChR [20] and characterized the corresponding aspartate to cysteine, and the histidine mutations as the aspartate in transmembrane helix 4 (TM4) were conserved in all three channelrhodopsins (Appendix B). Chronos D173C, D173H, and *Ps*ChR D139H, D139C, all showed an increased expression level, compared to their wild type (Figure 1a). All mutants also showed increased light-sensitivities along with prolonged off kinetics (Figure 2 and Table 1). Chronos D173C, *Ps*ChR D139H and *Ps*ChR D139C showed dramatically increased photocurrents while the Chronos D173H was similar to the wild type Chronos (Figure 1 and Table 1). Among these variants, *Ps*ChR D139C was the most light-sensitive with an effective light power density (LPD) for 50% photocurrent (EPD_{50}) of ~ 3.2 μW/mm^2, which was ~ 250 times more sensitive than *Ps*ChR (Figure 2 and Table 1). The EPD_{50} for XXL was ~ 5.4 μW/mm^2, which was ~ 130 times more sensitive than ChR2 (Figure 2 and Table 1).

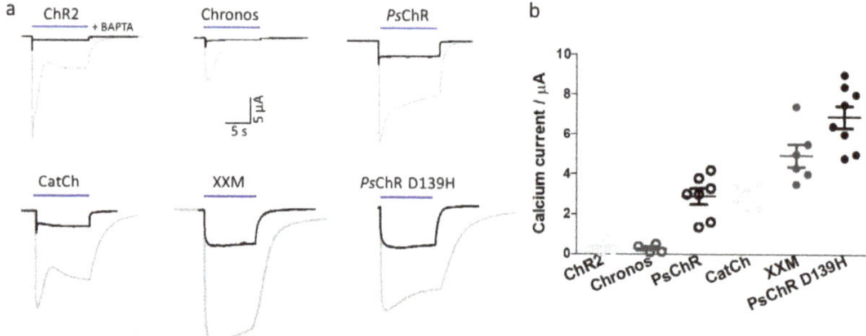

Figure 5. Calcium permeabilities of selected channelrhodopsin variants. (**a**) Photocurrent traces of different channelrhodopsins before (grey) and after (black) 1,2-bis(o-aminophenoxy)ethane-N,N,N′,N′-tetraacetic acid (BAPTA) injection. Measurements were done in 80 mM $CaCl_2$ pH 9.0 at −100mV. Blue bars indicate light illumination. (**b**) Comparison of calcium permeability of ChR2 (light grey), Chronos (dark grey), *Ps*ChR (black), CatCh (light grey), XXM (dark grey), and *Ps*ChR D139H (black). All data points were plotted in the figure and mean ± SEM was indicated.

Table 1. Basic properties of ChR2, *Ps*ChR, and Chronos variants.

	Expression *	I_s	Closing Time (ms) τ_1 (%‡)	Closing Time (ms) τ_2 (%‡)	EPD_{50} (μW/mm²)
ChR2	1× †	1× †	7.5 ± 0.5 (>98)	-	710
XXM	2.7 ×	30 ×	80 ± 5.5 (8)	1100 ± 110 (92)	90
XXL	3 ×	48 ×	-	71000 ± 2900 (>96)	5.4
Chronos	1 × †	1 × †	2.8 ± 0.3 (>99)	-	500
Chronos D173H	2.2 ×	1.2 ×	1400 ± 160 (56)	-	84
Chronos D173C	2.6 ×	10 ×	22 ± 1.3 (85)	130 ± 16 (15)	190
*Ps*ChR	1 × †	1 × †	8.5 ± 0.6 (>98)	-	810
*Ps*ChR D139H	2.2 ×	6 ×	37 ± 2.7 (14)	810 ± 54 (86)	160
*Ps*ChR D139C	3.1 ×	11 ×	3.4 ± 0.8 (4)	74000 ± 4300 (96)	3.2

†, expression or photocurrent of corresponding wild type was normalized as 1×. Is, stationary (plateau) current. ‡, as most ChRs exhibit biphasic off-kinetics which comprised a fast and a slow component, here the % indicated the percentage of the amplitude of the fast (τ_1) or slow (τ_2) component to the whole photocurrent. Data are shown as mean ± SEM, n = 4-5. Values are presented as approximates. *, expression level was calculated from the data of fluorescence value in Appendix A Figure A1.

Table 2. Ion selectivity of ChR2, *Ps*ChR, and Chronos variants.

	Reversal Potential Shift (mV)				Permeability Ratio		
	Na^+	H^+ †	K^+	H^+ ‡	Na^+/H^+	K^+/H^+	Na^+/K^+
ChR2	24 ± 0.6	20 ± 0.7	21 ± 0.6	22 ± 0.8	3.1×10^{-7}	2.5×10^{-7}	1.2
XXM	49 ± 1.5	8.2 ± 0.6	32 ± 1.0	11 ± 1.1	12×10^{-7}	5×10^{-7}	2.2
XXL	30 ± 2	10 ± 0.8	27 ± 0.8	10 ± 0.4	4.5×10^{-7}	3.8×10^{-7}	1.2
*Ps*ChR	58 ± 2.8	8.8 ± 5.5	54 ± 5.7	20 ± 1.3	18×10^{-7}	16×10^{-7}	1.2
*Ps*ChR D139H	91 ± 2.2	5.5 ± 0.6	65 ± 5.5	11 ± 0.6	90×10^{-7}	26×10^{-7}	3.5
*Ps*ChR D139C	51 ± 2.1	5 ± 0.4	40 ± 1.7	6.0 ± 0.6	13×10^{-7}	8×10^{-7}	1.7
Chronos	19 ± 0.4	21 ± 1.1	3.8 ± 0.6	53 ± 1.8	2.3×10^{-7}	0.33×10^{-7}	7.1
Chronos D173H	21 ± 2.4	14 ± 1.1	3.3 ± 0.6	38 ± 3.6	2.6×10^{-7}	0.28×10^{-7}	9.4
Chronos D173C	31 ± 2.4	9.5 ± 0.6	7.2 ± 0.7	37 ± 2.6	5×10^{-7}	0.67×10^{-7}	7.3

Reversal potentials and permeability ratio were determined from stationary currents in the indicated solution. Values represent mean ± SD, *n* = 4–6. Values without SD are presented as approximates. †, with the existence of 120 mM Na^+. ‡, with the existence of 120 mM K^+.

3.2. Mutation of the Aspartate in TM4 Influences the Na^+ Permeability

The potential influence on ion selectivity by mutating ChR2 D156 was not reported. To investigate this, we measured the photocurrent at different potentials and calculated the reversal potential shift of these mutants, when systematically changing bath solutions with different pH and Na^+ or K^+ concentrations.

ChR2 is a non-selective cation channel which is mostly permeable to H^+ [2]. Both changing extracellular pH from 7.6 to 9.6 and changing extracellular Na^+ concentration from 120 mM to 1 mM altered the ChR2 reversal potential (Figure 3). The ChR2 permeability ratio of Na^+ to H^+ (P_{Na+}/P_{H+}) was determined as $P_{Na+}/P_{H+} = 3.1\times10^{-7}$ (Table 2). Interestingly, D156H (XXM) influenced the Na^+ permeability and increased the P_{Na+}/P_{H+} four times, while D156C (XXL) changed the P_{Na+}/P_{H+} only slightly (Table 2).

Chronos is more permeable to H^+ than ChR2 with a $P_{Na+}/P_{H+} = 2.3\times10^{-7}$ (Table 2). The D173H mutation did not obviously change this, and D173C increased the P_{Na+}/P_{H+} slightly to 5×10^{-7} (Figure 3 and Table 2). *Ps*ChR was reported to be highly Na^+ conductive [20]. Changing the outside Na^+ concentration from 120 mM to 1 mM greatly influenced its photocurrent. The inward photocurrent was nearly abolished at 1 mM Na^+ pH 7.6, and we determined the *Ps*ChR P_{Na+}/P_{H+} to be 18×10^{-7}, which was even higher than that of XXM (Figure 3 and Table 2). The D139H mutation increased the P_{Na+}/P_{H+} even five-fold more, to 90×10^{-7}, while the D139C mutation decreased the P_{Na+}/P_{H+} slightly (Figure 3 and Table 2).

Among the tested constructs, *Ps*ChR D139H was the most Na^+ permeable channelrhodopsin with a large photocurrent and, to our knowledge, the most Na^+-permeable channelrhodopsin ever reported.

3.3. Mutation of the Aspartate in TM4 Influences the K^+ Permeability

As tools for light-induced depolarization, ideal cation-permeable channelrhodopsins should be more Na^+ conductive and less K^+ conductive, because K^+ efflux across the plasma membrane would lead to a more hyperpolarized membrane potential. To test the potential influences on K^+ permeability of different mutations, we measured photocurrents and calculated the reversal potential shift of these mutants when systematically changing bath solutions from 120 mM K^+ to 1 mM K^+ in comparison to changing pH from 7.6 to 9.6.

ChR2 had a slightly weaker K^+ conductance in comparison to Na^+ with a $P_{Na+}/P_{K+} = 1.2$. XXL increased the Na^+ and K^+ permeability slightly and equally. XXM increased the Na^+ permeability more than that for K^+, and the P_{Na+}/P_{K+} of XXM reached 2.2 (Figure 4 and Table 2). *Ps*ChR showed a higher Na^+ permeability, together with an enhanced K^+ permeability in comparison to that for H^+, with a similar P_{Na+}/P_{K+} as ChR2 (Figure 4 and Table 2). Interestingly, the D139H mutation increased the Na^+ permeability five-fold, while changing the K^+ permeability only 1.6-fold, thus the P_{Na+}/P_{K+} of *Ps*ChR D139H increased to 3.5 (Figure 4 and Table 2). The increased P_{Na+}/P_{K+} makes *Ps*ChR D139H even more suitable as a depolarization tool.

Chronos, Chronos D173H, and Chronos D173C had much lower K^+ permeability and the highest P_{Na+}/P_{K+} value among the tested constructs (Figure 4 and Table 2). However, the H^+ permeability was the highest for all Chronos variants (Table 2).

3.4. Mutation of the Aspartate in TM4 Influences the Ca^{2+} Permeability

As obvious impacts of mutation of the conserved aspartate in TM4 on ion selectivity were observed, we further compared the Ca^{2+} permeability of these mutants, considering the importance of Ca^{2+} in biological systems. Due to the existence of Ca^{2+}-activated chloride channels in *Xenopus* oocytes [23], BAPTA was injected into the oocyte to a final concentration of ~10 mM, to block the Ca^{2+}-induced chloride current (Figure 5). Then the photocurrents at −100 mV were measured in outside solution containing 80 mM $CaCl_2$ at pH 9.0. At −100 mV and pH 9, no net H^+ current could be observed and the inward photocurrent was then only from the Ca^{2+} influx.

ChR2 showed a robust composite photocurrent with 80 mM $CaCl_2$ at pH 9.0 and −100 mV, which was dramatically reduced to the pure Ca^{2+} current after injection of 10 mM BAPTA (Figure 5a), as reported previously [2]. Both Chronos and ChR2 showed small Ca^{2+} photocurrents (Figure 5a). ChR2 L132C (CatCh) showed an increased Ca^{2+} photocurrent, compared to ChR2 (Figure 5), as previously reported [8]. Astonishingly, XXM also showed an increased Ca^{2+} photocurrent, even higher than that of CatCh (Figure 5). *Ps*ChR D139H showed the highest Ca^{2+} photocurrent, which on average was more than two times higher than that of CatCh (Figure 5b).

4. Discussion

Channelrhodopsins, originating from different organisms, show quite different properties with respect to kinetics, action spectrum, and ion selectivity. Such changes can also be engineered by point mutations. In this study we compared the properties of ChR2, Chronos [9], *Ps*ChR [20], and their corresponding mutants of the aspartate in TM4 (DC gate aspartate).

Generally, the aspartate to histidine or cysteine mutations of the three channelrhodopsins increased the expression level (probably because the mutant became more stable against degradation [21]) and slowed the closing kinetics. Nearly all mutants showed a much-increased photocurrent, probably because of a much-prolonged open state or enhanced single channel conductance, with only Chronos D173H as an exception.

The tools with slowed kinetics are unfavorable for ultra-fast multiple stimulation but preferred for experiments which require low light and longtime stimulation. The prolonged open times were accompanied by elevated light sensitivities. Among the tested constructs, *Ps*ChR D139C and XXL became ~ 220 times and ~ 130 times more sensitive than ChR2. If slow closing would have not been a problem nor even desired, the more light-sensitive channelrhodopsins would have been ideal for efficient deep brain stimulation with infrared light via upconversion nanoparticles (UCNPs) [24]. These tools need to be further tested in mammalian systems for a broader field application.

Furthermore, we investigated the influence of mutation of the aspartate in TM4 on ion selectivity. We found that aspartate to histidine mutation of ChR2 and *Ps*ChR increased the Na^+ and Ca^{2+} permeability dramatically. To test the Ca^{2+} current, we used BAPTA to block the Ca^{2+}-activated endogenous chloride channels of oocytes. The fast Ca^{2+} chelator BAPTA may have been altering the ion currents in more ways [25]. However, as we could see from the kinetics in Figure 5a that the Cl^- current (which shows a slower off kinetics) was well-blocked. Then we could reliably compare only the photocurrent of our channelrhodopsins.

With the large photocurrent, increased Na^+ permeability, and bigger Ca^{2+} current, *Ps*ChR D139H is a novel powerful optogenetic tool for depolarization and Ca^{2+} manipulation. Channelrhodopsins with higher Ca^{2+} currents have the advantage of being "direct" light-gated Ca^{2+} channels, in contrast to the highly Ca^{2+}-conductive CNG (cyclic nucleotide-gated) channels which became light-gated channels when fused with bPAC (photoactivated adenylyl cyclase) [26].

In summary, we found that mutating the conserved aspartate in TM4 influenced not only the expression level and kinetics of channel closing but also the ion selectivity; with appropriate mutations, we provided novel optogenetic tools with superior photocurrent amplitudes and high Na^+ and Ca^{2+} conductance.

Author Contributions: Conceptualization, S.G. and G.N.; methodology, X.D., S.G. and G.N.; software, X.D. and S.G.; validation, X.D., S.G. and G.N.; formal analysis, X.D. and S.G.; investigation, X.D. and S.G.; resources, X.D., S.G. and G.N.; data curation, X.D. and S.G.; writing—original draft preparation, S.G.; writing—review and editing, X.D., S.G. and G.N.; visualization, X.D. and S.G.; supervision, S.G. and G.N.; project administration, G.N.; funding acquisition, G.N.

Funding: This research was funded by grants from the German Research Foundation to GN (TRR 166/A03 and TR 240/A04). GN acknowledges support provided by the Prix-Louis-Jeantet.

Acknowledgments: We are grateful to Shang Yang for help with some of the cloning work. This publication was funded by the German Research Foundation (DFG) and the University of Wuerzburg in the funding program Open Access Publishing.

Conflicts of Interest: The authors declare no conflict of interest.

Appendix A

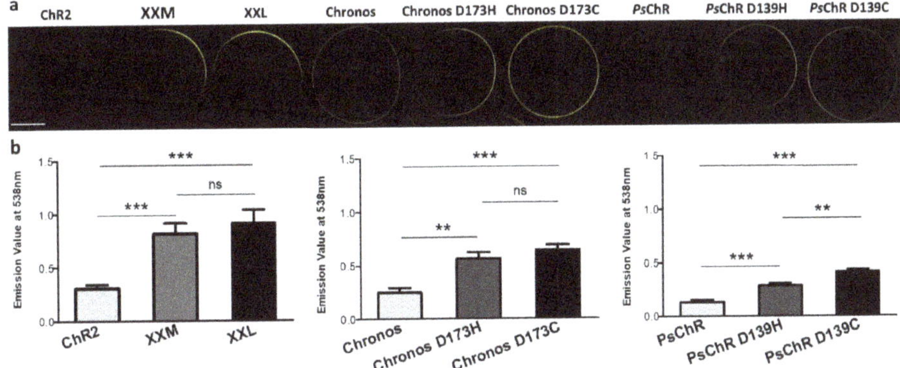

Figure A1. Expression level of ChR2, *Ps*ChR, and Chronos variants in *Xenopus* oocyte. (**a**) Representative confocal images of all the constructs, scale bar = 500 µm. (**b**) Yellow fluorescent protein (YFP) fluorescence emission values from oocytes expressing different channelrhodopsins. Data was shown as mean ± SEM, n = 5–6. Pictures and fluorescence emission values were taken and measured 2 days after 20 ng cRNA injection.

Appendix B

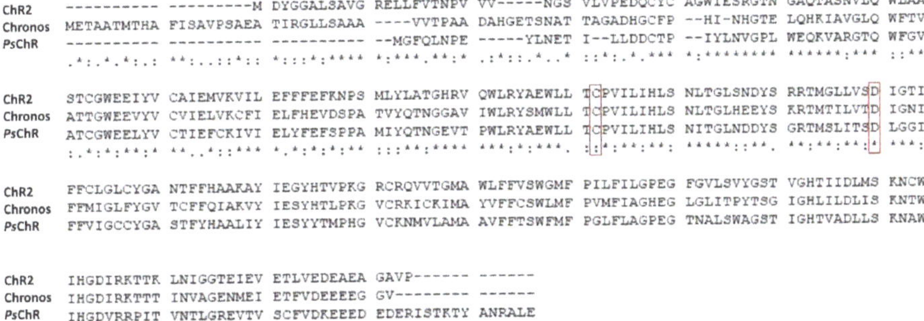

Figure A2. Sequence alignment of ChR2, Chronos, and *Ps*ChR. Conserved cysteine and aspartate of the DC gate were marked in the red box.

References

1. Nagel, G.; Ollig, D.; Fuhrmann, M.; Kateriya, S.; Mustl, A.M.; Bamberg, E.; Hegemann, P. Channelrhodopsin-1: A light-gated proton channel in green algae. *Science* **2002**, *296*, 2395–2398. [CrossRef]
2. Nagel, G.; Szellas, T.; Huhn, W.; Kateriya, S.; Adeishvili, N.; Berthold, P.; Ollig, D.; Hegemann, P.; Bamberg, E. Channelrhodopsin-2, a directly light-gated cation-selective membrane channel. *Proc. Natl. Acad. Sci. USA* **2003**, *100*, 13940–13945. [CrossRef] [PubMed]
3. Boyden, E.S.; Zhang, F.; Bamberg, E.; Nagel, G.; Deisseroth, K. Millisecond-timescale, genetically targeted optical control of neural activity. *Nat. Neurosci.* **2005**, *8*, 1263–1268. [CrossRef] [PubMed]
4. Li, X.; Gutierrez, D.V.; Hanson, M.G.; Han, J.; Mark, M.D.; Chiel, H.; Hegemann, P.; Landmesser, L.T.; Herlitze, S. Fast noninvasive activation and inhibition of neural and network activity by vertebrate rhodopsin and green algae channelrhodopsin. *Proc. Natl. Acad. Sci. USA* **2005**, *102*, 17816–17821. [CrossRef] [PubMed]

5. Nagel, G.; Brauner, M.; Liewald, J.F.; Adeishvili, N.; Bamberg, E.; Gottschalk, A. Light activation of channelrhodopsin-2 in excitable cells of *Caenorhabditis elegans* triggers rapid behavioral responses. *Curr Biol.* **2005**, *15*, 2279–2284. [CrossRef] [PubMed]
6. Bi, A.D.; Cui, J.J.; Ma, Y.P.; Olshevskaya, E.; Pu, M.L.; Dizhoor, A.M.; Pan, Z.H. Ectopic expression of a microbial-type rhodopsin restores visual responses in mice with photoreceptor degeneration. *Neuron.* **2006**, *50*, 23–33. [CrossRef] [PubMed]
7. Ishizuka, T.; Kakuda, M.; Araki, R.; Yawo, H. Kinetic evaluation of photosensitivity in genetically engineered neurons expressing green algae light-gated channels. *Neurosci. Res.* **2006**, *54*, 85–94. [CrossRef]
8. Kleinlogel, S.; Feldbauer, K.; Dempski, R.E.; Fotis, H.; Wood, P.G.; Bamann, C.; Bamberg, E. Ultra light-sensitive and fast neuronal activation with the Ca^{2+}-permeable channelrhodopsin CatCh. *Nat. Neurosci.* **2011**, *14*, 513–518. [CrossRef]
9. Klapoetke, N.C.; Murata, Y.; Kim, S.S.; Pulver, S.R.; Birdsey-Benson, A.; Cho, Y.K.; Morimoto, T.K.; Chuong, A.S.; Carpenter, E.J.; Tian, Z.J.; et al. Independent optical excitation of distinct neural populations. *Nat. Methods* **2014**, *11*, 972. [CrossRef]
10. Wietek, J.; Wiegert, J.S.; Adeishvili, N.; Schneider, F.; Watanabe, H.; Tsunoda, S.P.; Vogt, A.; Elstner, M.; Oertner, T.G.; Hegemann, P. Conversion of Channelrhodopsin into a Light-Gated Chloride Channel. *Sci.* **2014**, *344*, 409–412. [CrossRef]
11. Govorunova, E.G.; Sineshchekov, O.A.; Janz, R.; Liu, X.Q.; Spudich, J.L. Natural light-gated anion channels: A family of microbial rhodopsins for advanced optogenetics. *Sci.* **2015**, *349*, 647–650. [CrossRef] [PubMed]
12. Berndt, A.; Yizhar, O.; Gunaydin, L.A.; Hegemann, P.; Deisseroth, K. Bi-stable neural state switches. *Nat. Neurosci.* **2009**, *12*, 229–234. [CrossRef]
13. Radu, I.; Bamann, C.; Nack, M.; Nagel, G.; Bamberg, E.; Heberle, J. Conformational changes of channelrhodopsin-2. *J. Am. Chem. Soc.* **2009**, *131*, 7313–7319. [CrossRef]
14. Bamann, C.; Gueta, R.; Kleinlogel, S.; Nagel, G.; Bamberg, E. Structural Guidance of the Photocycle of Channelrhodopsin-2 by an Interhelical Hydrogen Bond. *Biochem.* **2010**, *49*, 267–278. [CrossRef] [PubMed]
15. Nack, M.; Radu, I.; Gossing, M.; Bamann, C.; Bamberg, E.; von Mollard, G.F.; Heberle, J. The DC gate in Channelrhodopsin-2: crucial hydrogen bonding interaction between C128 and D156. *Photochem. Photobiol. Sci.* **2010**, *9*, 194–198. [CrossRef] [PubMed]
16. Kato, H.E.; Zhang, F.; Yizhar, O.; Ramakrishnan, C.; Nishizawa, T.; Hirata, K.; Ito, J.; Aita, Y.; Tsukazaki, T.; Hayashi, S.; et al. Crystal structure of the channelrhodopsin light-gated cation channel. *Nat.* **2012**, *482*, 369–374. [CrossRef] [PubMed]
17. Volkov, O.; Kovalev, K.; Polovinkin, V.; Borshchevskiy, V.; Bamann, C.; Astashkin, R.; Marin, E.; Popov, A.; Balandin, T.; Willbold, D.; et al. Structural insights into ion conduction by channelrhodopsin 2. *Sci.* **2017**, *358*, eaan8862. [CrossRef]
18. Dawydow, A.; Gueta, R.; Ljaschenko, D.; Ullrich, S.; Hermann, M.; Ehmann, N.; Gao, S.; Fiala, A.; Langenhan, T.; Nagel, G.; et al. Channelrhodopsin-2-XXL, a powerful optogenetic tool for low-light applications. *Proc. Natl. Acad. Sci. USA* **2014**, *111*, 13972–13977. [CrossRef]
19. Scholz, N.; Guan, C.; Nieberler, M.; Grotemeyer, A.; Maiellaro, I.; Gao, S.; Beck, S.; Pawlak, M.; Sauer, M.; Asan, E.; et al. Mechano-dependent signaling by Latrophilin/CIRL quenches cAMP in proprioceptive neurons. *eLife* **2017**, *6*, e28360. [CrossRef]
20. Govorunova, E.G.; Sineshchekov, O.A.; Li, H.; Janz, R.; Spudich, J.L. Characterization of a highly efficient blue-shifted channelrhodopsin from the marine alga *Platymonas subcordiformis*. *J. Biol. Chem.* **2013**, *288*, 29911–29922. [CrossRef]
21. Ullrich, S.; Gueta, R.; Nagel, G. Degradation of channelopsin-2 in the absence of retinal and degradation resistance in certain mutants. *Biol. Chem.* **2013**, *394*, 271–280. [CrossRef] [PubMed]
22. Gradinaru, V.; Zhang, F.; Ramakrishnan, C.; Mattis, J.; Prakash, R.; Diester, I.; Goshen, I.; Thompson, K.R.; Deisseroth, K. Molecular and cellular approaches for diversifying and extending optogenetics. *Cell* **2010**, *141*, 154–165. [CrossRef] [PubMed]
23. Boton, R.; Dascal, N.; Gillo, B.; Lass, Y. Two calcium-activated chloride conductances in *Xenopus laevis* oocytes permeabilized with the ionophore A23187. *J. Physiol.* **1989**, *408*, 511–534. [CrossRef] [PubMed]
24. Chen, S.; Weitemier, A.Z.; Zeng, X.; He, L.; Wang, X.; Tao, Y.; Huang, A.J.Y.; Hashimotodani, Y.; Kano, M.; Iwasaki, H.; et al. Near-infrared deep brain stimulation via upconversion nanoparticle-mediated optogenetics. *Sci.* **2018**, *359*, 679–684. [CrossRef] [PubMed]

25. Bootman, M.D.; Allman, S.; Rietdorf, K.; Bultynck, G. Deleterious effects of calcium indicators within cells; an inconvenient truth. *Cell Calcium* **2018**, *73*, 82–87. [CrossRef] [PubMed]
26. Beck, S.; Yu-Strzelczyk, J.; Pauls, D.; Constantin, O.M.; Gee, C.E.; Ehmann, N.; Kittel, R.J.; Nagel, G.; Gao, S. Synthetic Light-Activated Ion Channels for Optogenetic Activation and Inhibition. *Front. Neurosci.* **2018**, *12*, 643. [CrossRef] [PubMed]

Article

Channelrhodopsin-2 Function is Modulated by Residual Hydrophobic Mismatch with the Surrounding Lipid Environment

Ryan Richards [1,2,*,†], Sayan Mondal [3,4,†], Harel Weinstein [3,5] and Robert E. Dempski [1]

1 Department of Chemistry and Biochemistry, Worcester Polytechnic Institute, Worcester, MA 01609, USA
2 Program in Systems Biology, University of Massachusetts Medical School, Worcester, MA 01655, USA
3 Department of Physiology and Biophysics, Weill Cornell Medical College of Cornell University, New York, NY 10065, USA
4 Schrödinger Inc., New York, NY 10036, USA
5 Institute for Computational Biomedicine, Weill Cornell Medical College, Cornell University, New York, NY 10065, USA
* Correspondence: ryan.richards@umassmed.edu; Tel.: +774-455-3870
† These authors contributed equally to this manuscript.

Received: 28 April 2019; Accepted: 26 June 2019; Published: 30 June 2019

Featured Application: This report highlights the importance of the biological membrane in channelrhodopsin-2 function and provides a framework for targeted engineering of ChR2 constructs for specific membrane compositions.

Abstract: Channelrhodopsin-2 (ChR2) is a light-gated ion channel that conducts cations of multiple valencies down the electrochemical gradient. This light-gated property has made ChR2 a popular tool in the field of optogenetics, allowing for the spatial and temporal control of excitable cells with light. A central aspect of protein function is the interaction with the surrounding lipid environment. To further explore these membrane-protein interactions, we demonstrate the role of residual hydrophobic mismatch (RHM) as a mechanistically important component of ChR2 function. We combined computational and functional experiments to understand how RHM between the lipid environment and ChR2 alters the structural and biophysical properties of the channel. Analysis of our results revealed significant RHM at the intracellular/lipid interface of ChR2 from a triad of residues. The resulting energy penalty is substantial and can be lowered via mutagenesis to evaluate the functional effects of this change in lipid-protein interaction energy. The experimental measurement of channel stability, conductance and selectivity resulting from the reduction of the RHM energy penalty showed changes in progressive H^+ permeability, kinetics and open-state stability, suggesting how the modulation of ChR2 by the surrounding lipid membrane can play an important biological role and contribute to the design of targeted optogenetic constructs for specific cell types.

Keywords: *Chlamydomonas reinhardtii*; ion channel; optogenetics; electrophysiology; molecular dynamics simulations; membrane-protein interaction; energy of membrane deformation; CTMD method, residual hydrophobic mismatch

1. Introduction

The eyespot region of the green algae Chlamydomonas reinhardtii contains two photoreceptor proteins, channelrhodopsin-1 (ChR1) and channelrhodopsin-2 (ChR2), which are implicated in the first committed step of the phototactic response [1–3]. ChR2 is a light-activated ion channel that conducts cations of multiple valencies down the electrochemical gradient [1,2]. ChR2 is part of the microbial-rhodopsin family of membrane proteins found in archaea, algae and bacteria [4]. Its structure follows the common architecture of seven-helix transmembrane domains (TMD) that houses all-trans retinal bound through a protonated Schiff base to a completely conserved lysine residue on TM 7 (K257 in ChR2). Photoactivation of ChR2 with blue light (λ_{max} = 470 nm) isomerizes retinal to the 13-cis isoform to begin the photocycle reaction that leads to ion conductance [2]. The functional properties of ChR2 have made it a popular tool in the field of optogenetics in which excitable cells expressing photoactivatable ion channels are controlled by light. ChR2 activation can rapidly depolarize the membrane of excitable cells and induce neuronal spiking of action potentials with high temporal and spatial resolution [5].

ChR2 photocurrents rise to an initial transient peak (I_p), which decays rapidly to a steady-state value (I_{ss}) in a process known as inactivation [6]. Once illumination ends, I_{ss} decays biexponentially to baseline, indicative of a two-step process [7]. Interestingly, even after long periods of recovery in the dark, ChR2 exhibits reduced peak currents on subsequent light pulses, which is referred to as desensitization [8]. There have been multiple attempts to generate a photocycle model that accurately recreates ChR2 photocurrents. Several kinetic three and four-state models have been developed to reproduce and quantify ChR2 photocurrents [9,10]. The current models consist of two closed states and two open states (Figure 1) [11–13]. Here, the ground state C1 rapidly transitions to an initial high conducting O1 state. Under continuous illumination conditions, O1 decays to the lower-conducting O2 state. Once the light is turned off, ion conductance ends and O2 moves to the desensitized closed state C2. During sufficient time in the dark, C2 thermally converts back to C1.

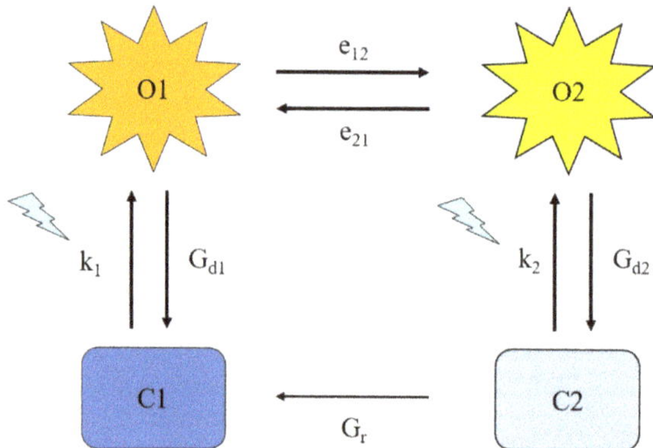

Figure 1. Channelrhodopsin-2 (ChR2) four-state model. Four-state kinetic model of ChR2 with two conducting (O1 and O2) and two non-conducting (C1 and C2) states. Each transition is represented by the corresponding rate parameter used in the kinetic model. The blue arrows represent light-assisted transitions.

The ChR2 kinetic model is defined by the distribution of conformational states (O1, O2, C1, C2) with corresponding rates of transition and the ion conductance properties of these states. Molecular detail relating the timing of conformational changes during the photocycle to the resulting ion conductance is required for a better understanding that can lead to predictive modeling of ChR2 function. Indeed, compared to microbial-rhodopsin pumps, ChR2 undergoes much larger conformational changes in backbone structure prior to ion conductance [14–16]. The analysis of conformational rearrangements of the ChR2 molecule from spectroscopic measurements, spin-labeled electron paramagnetic resonance (EPR) experiments and molecular dynamic (MD) simulations has indicated activation-related movement of transmembrane segment (TM) 2 away from the bundle by over 3 Å [16,17]. This movement is also sensed by TM7 through an interhelical hydrogen-bonding network. The resulting reorientation of the gating residues E90, N258 and Y70 enables water penetration to the pore and subsequent ion conductance [15].

Given the substantial structural rearrangements of transmembrane segments (TMs) in the TMD associated with ChR2 function, it becomes important to evaluate the role and contribution of the local membrane environment in affecting channel mechanism. Such membrane-protein interactions have been shown to play central roles in the functional mechanisms of membrane proteins—from G-protein-coupled receptors (GPCRs) [18–21] to transporters [22–24] and channels [25,26]—but have not been addressed in the analysis of ChR2 function. Since the physical properties of membranes, such as thickness, curvature, compression modulus and specific composition have all been shown to affect the structure and function of the numerous membrane proteins studied, it seems essential to explore the role of membrane interactions of ChR2 in order to connect expected functional properties to diverse environments in various cells.

The membrane-mediated regulation of proteins has been addressed quantitatively in terms of hydrophobic mismatch [21,22,27]. However, as the TM helix bundle of membrane proteins is highly asymmetric (with TM segments of dissimilar length and orientation), the evaluation of the mismatch is a complex task. Moreover, the conformational rearrangements associated with the various states of these proteins (active, inactive, desensitized) engender dynamic interaction with the lipid environment. Consequently, the hydrophobic mismatch that occurs when the membrane spanning regions of proteins do not position hydrophobic/hydrophilic regions in alignment with their counterparts in the bilayer is dynamic in nature, following the interfaces created in the protein states by deforming the bilayer so that the thickness of the hydrophobic portion of the lipid bilayer can match the hydrophobic core of the protein. This matching prevents the exposure of nonpolar (or polar) residues to the inappropriate part of the membrane and minimizes the energetic cost. If the adaptation of the membrane does not fully compensate the hydrophobic core, a residual hydrophobic mismatch (RHM) is established [19], which incurs additional energy costs [22]. In simple channels, such as the single, helical TMs gramicidin A channel, the hydrophobic mismatch can be remedied mostly by simple membrane deformations [28]. However, in more complex channels composed of multiple TMs the membrane adaptation is challenged by the radially asymmetric hydrophobicity of the interface surface, which also changes with the functional or oligomerization state of the protein.

The recently developed computational approach—Continuum Theory-Molecular Dynamics (CTMD)—has been shown to account quantitatively for the effects of complex lipid deformations that occur with multiple TM helices [27]. Once the membrane deformation profile is established with CTMD for each conformational state of interest, the method also quantifies the solvent accessible area of unfavorably exposed residues due to RHM and calculates the resulting energy penalty [18,22]. Because the CTMD method calculates the energy components in a specific structural context that identifies the residues producing the largest energy penalty, the results predict specific changes that can be introduced by mutagenesis to reduce (or increase) the RHM and show the dependence of RHM on the composition-dependent properties of the bilayer (e.g., thickness). Thus, the methodology can serve prospectively to identify residues critical to hydrophobic mismatch in different lipid environments and on this basis predict specific mutations that would increase/decrease the compatibility between the protein and the specific bilayer, thereby guiding mutagenesis experiments and choices of preferred membrane environments for activity.

ChR2 has been successfully expressed in functionally diverse cell lines, each with unique membrane compositions [5,29–31]. To further explore these membrane-protein interactions, we demonstrate the role of RHM as a mechanistically important component of ChR2 function. Through a combination of MD simulations, CTMD calculations and functional measurements, we have quantified the key contributions to RHM of residues within TM1 and TM7. For several decades, hydrophobic mismatch has been proposed to modulate TM protein function. This may be the first report describing the identification of particular residues and corresponding mutations central to the role of hydrophobic mismatch that was also followed by experimental validation for a multi-TM protein. Notably, our results provide a physical premise for previously observed effects on pore diameter by the V269S ChR2 mutation reported by us [32]. Analysis of our results highlights the importance of the lipid membrane in modulating the functional properties of ChR2 during optogenetic applications.

2. Materials and Methods

The membrane-protein interactions of the ChR2 protein were quantified with the Continuum-Molecular Dynamics (CTMD) hybrid approach using a protocol described in detail previously [22–27]. The all atom MD simulations were carried out following in detail the protocols that served in our simulations of membrane proteins (e.g., see References [22,33–35]), employing NAMD software and the CHARMM36 force field for the protein and the membrane. ChR2 was modeled on the structure of the C1C2 chimera [13,36,37]. Alignment with the X-ray structure of ChR2 was performed using the MultiSeq package in VMD [38,39]. The model was embedded in a patch of 1-palmitoyl-2-oleoyl-sn-glycero-3-phosphocholine (POPC) lipid bilayer of dimensions ~100 Å by 100 Å. The length of each simulation of the resulting system was at least 200ns.

Quantification of the residual hydrophobic mismatch was achieved as previously described [22,27]. In brief, the surface area $SA_{res,i}$ of the ith residue participating in unfavorable hydrophobic-polar interactions (i.e., when the residue's hydrophobicity does not match that of its environment) was calculated using modified residue-specific solvent accessible surface area (SASA). For hydrophobic residues, the RHM is $SA_{mem,i}$, where $SA_{mem,i}$ = SASA calculated for the solute comprising the protein and the hydrophobic core of the lipid bilayer (C2-C2). For polar residues, the RHM is calculated as $SA_{res,i} = SA_{prot,i} - SA_{mem,i}$, where polar residues are on the surface of the protein and within the lipid bilayer hydrophobic core and $SA_{prot,i}$ = SASA with the solute comprising the protein only. An exception is made for tryptophan, which is favorably located at the interface [40].

The energy penalty is directly proportional to $SA_{res,i}$ and the energy penalty for the Nth TM (N = 1 to 7) is given by:

$$\Delta G_{res} = \sum_{i=1}^{N_{TM}} \Delta G_{res,i} \sim \sum_{i=1}^{N_{TM}} \sigma_{res} SA_{res,i} \tag{1}$$

where the constant of proportionality σ_{res} is taken as 0.028 kcal/(mol. Å^2) and N_{TM} is the number of membrane-exposed residues in the TM [41].

As described, neither arginine nor lysine are penalized when they are located close to the lipid headgroups [42]. In addition, serine and threonine exposed to the hydrophobic membrane interface can form hydrogen bonding with the helix backbone and are therefore not penalized [43].

SASA are computed from all-atom molecular dynamics simulations (described below), with a probe radius of 1.4 Å and using the *g sas* utility of the GROMACS software [44].

2.1. Membrane Deformation

Membrane deformation can be described as a function of the local bilayer thickness $d(x,y)$ and the equation:

$$\mu(x,y) = \frac{1}{2}(d(x,y) - d_0) \tag{2}$$

where d_0 is the bulk thickness of the bilayer away from the protein. The trajectory is centered at that protein and the time-averaged $d(x,y)$ is computed by fitting a 2Å x 2Å rectangular grid to the phosphate beads over the course of the trajectory, followed by spatial smoothing. In this manner, the $d(x,y)$ can then be calculated around ChR2 from the MD trajectory. More complete details of this methodology can be found in Reference [27].

For the calculation of deformation free energies using the CTMD method, the membrane is considered an elastic continuum, where the deformation energy cost is the sum of the compression-extension, splay-distortion and surface tension components [28,45]. To obtain $\mu(x,y)$ the Euler-Lagrange equation is solved with no assumption in regard to radial symmetry and with specific boundary conditions on the protein-membrane boundary taken from the MD simulation results as described previously [27]. The solution to the Euler-Lagrange equation for the ChR2 systems yields the energy cost of the membrane deformation computed using the boundary conditions from the MD simulations.

2.2. Molecular Biology, mRNA Synthesis and Oocyte Injection

The gene encoding for a truncated ChR2 (residues 1-308) with a C-terminal hemagglutinin tag was cloned into the pTLN vector between EcoRV and XbaI restriction sites [13,32,37,46]. All point mutations were created using the QuikChange site-directed mutagenesis (Agilent Genomics). Mutations were verified by full gene sequencing.

Oocytes were extracted from female *Xenopus laevis* frogs by partial ovariectomy. Oocytes were digested in ORI (-) buffer (90 mM NaCl, 2 mM KCl, 5 mM MOPS; pH 7.4) supplemented with 3 mg/mL collagenase type II at 17 °C with gentle shaking. After digestion, oocytes were washed with copious amounts of ORI (+) buffer (90 mM NaCl, 2 mM KCl, 2 mM $CaCl_2$, 5 mM MOPS; pH 7.4). ChR2 mRNA was synthesized in vitro using the SP6 mMessage mMachine kit (Agilent, Santa Clara, CA, USA). The mRNA was diluted to 1 ug/uL in DEPC-treated water. A volume of 50 nL mRNA was injected into each oocyte (50 ng total). Post-injection, oocytes were stored in 5 mL of ORI (+) supplemented with 1.5 μM all-trans retinal and 1 mg/mL gentamycin at 17 °C in the dark for 3–4 days.

2.3. Electrophysiology

Borosilicate glass electrodes were pulled using a PC-10 puller (Narishge, Amityville, NY, USA). The microelectrodes were filled with 3 M KCl and had tip resistances between 0.5–1.5 MΩ. Voltage-clamping of oocytes was achieved through a Turbo-tec 03X amplifier connected to a 1440 A Axon Instruments digitizer (Molecular Devices, San Jose, CA, USA). Voltage protocols and current recording were controlled with Clampex software. ChR2 was activated via a 1 mm light guide connected to an Omicron LEDMOD V2 300 mW LED module with an emission wavelength of 470 nm. The membrane voltage was varied from −100 mV to +40 mV in 20 mV steps. Sodium solutions consisted of 100 mM NaCl, 2 mM $MgCl_2$, 0.1 $CaCl_2$, 5 mM MOPS/5 mM MES at pH 7.5 and 6.0, respectively. Reversal potentials were determined from current-voltage curves at I_p and I_{ss}. Kinetic parameters were determined by fitting photocurrent traces with monoexponetial (T_{decay}) or biexponential equations (T_{off}).

2.4. Four-state modeling of photocycle kinetics

The mathematical model describing ChR2 photocycle kinetics was developed based on previous four-state models (Figure 1) [11–13,45]. Transitions between the two closed (C1 and C2) and two open (O1 and O2) states can be described by the following rate equations:

$$\frac{dC1}{dt} = G_r \times O2 + G_{d1} \times O1 - k_1 \times C1 \tag{3}$$

$$\frac{dC2}{dt} = G_{d2} \times O2 - (k_1 + G_r) \times C2 \tag{4}$$

$$\frac{dO1}{dt} = k_1 \times C1 + e_{21} \times O2 - (G_{d1} + e_{12}) \times O1 \tag{5}$$

$$\frac{dO2}{dt} = k_2 \times C2 + e_{12} \times O1 - (G_{d2} + e_{21}) \times O2 \tag{6}$$

$$C1 + C2 + O1 + O2 = 1 \tag{7}$$

where C1, C2, O1 and O2 represent the population of each state, G_r is the thermal conversion of C2 → -C1, $G_{d1/2}$ are the back reactions O1(2) → C1(2), e_{12} is the O1 → O2 transition and e_{21} is the reverse reaction. The light-assisted transitions $k_{1/2}$ represents activation of C1 and C2 respectively and are described by:

$$k_{1/2} = \varepsilon 1/2\mathrm{Fp} \tag{8}$$

where ε is the quantum efficiency of retinal photon absorption in C1 or C2, F is the photon flux and p is the state-dependent activation function related to the non-instantaneous response of retinal to light described by References [47,48]:

$$\frac{dp}{dt} = (\mathrm{S_0}(\Theta) - \mathrm{p})/\mathrm{T_{ChR2}} \tag{9}$$

where $S_0(\Theta)$ is the sigmoidal function and T_{ChR2} is the time constant for activation. Finally, the solutions to the rate equations are used to determine the current (I) flowing through ChR2:

$$I = \mathrm{I_{max}}\ (\mathrm{O1} + \gamma \mathrm{O2}) \tag{10}$$

With γ representing the conductance ratio between the O2/O1 individual conductances g_2 and g_1 and I_{max} is $V_m x g_1$.

Photocurrent traces were imported into Mathworks MATLAB 2013. The parameters describing the photocycle (k_1, k_2, G_{d1}, G_{d2}, e_{12}, e_{21}, I_{max}, γ, G_r, T_{ChR2}) were optimized using the FMINCON function and used to fit experimental traces. Differential equations were solved using the built-in ode23 function. Minimization between the theoretical and experimental traces was done using a simple sum of squared error routine.

3. Results

3.1. Identification and Quantification of Residual Hydrophobic Mismatch

To quantify residual hydrophobic mismatch for ChR2 in the POPC membranes, we obtained an extended simulation trajectory from MD simulations of the ChR2 homology model based on the C1C2 structure, embedded in a POPC lipid bilayer. The ChR2 homology model aligns closely with the recently solved X-ray structure of ChR2 (Figure 2A) with a RMSD of 1.17 Å (Figure 2A) [39]. The most notable structural differences occur in the N-terminal domain and the loop between TM1 and TM2. The CTMD approach was then used as previously described [22,27] to calculate the consequences of protein-membrane interaction, including membrane deformation and the free energy cost of the RHM. The analysis revealed significant residual hydrophobic mismatch in WT ChR2 (Table 1). Although membrane deformation had alleviated some unfavorable interactions, residues located at the cytoplasmic interface of TMs 2, 6 and 7 remained exposed to either polar or hydrophobic inappropriate environments due to the local adjacency to residues with diverse properties that prevented further adaptation of the membrane (see References [22,27]) Consequently, Q73 on TM1 and Q210 on TM6 were found to face the hydrophobic interface of the membrane. This caused membrane thinning in the adjacent regions of both residues, leading to water penetration into the membrane and exposure of V269 to a polar environment (Figure 2B). Substantial RHM was also observed at N187 and V269. Notably, mutations at these positions have previously been shown to affect ChR2 function [32,49].

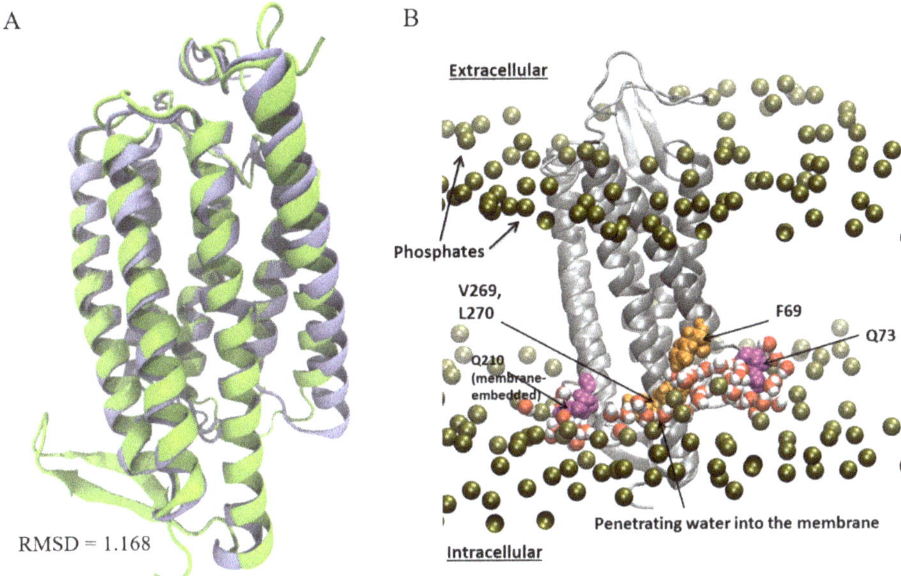

Figure 2. Homology model comparison and structural context of hydrophobic mismatch. (**A**) Structural alignment of the ChR2 homology model (lime) used for the calculations and the ChR2 X-ray structure (blue). (**B**) The averaged membrane deformation around ChR2 is shown on a snapshot of the protein, with the residues where residual hydrophobic mismatch (RHM) occurs highlighted. The membrane deformations are shown in terms of the time averaged phosphate surface around the centered protein and water penetration. RMSD: Root mean square deviation

Table 1. Percent change of residual hydrophobic mismatch for channelrhodopsin-2 when compared to wild type protein, calculated with respect to the RHM of the entire protein. The average energy cost of RHM for WT channelrhodopsin-2 and select mutant constructs embedded in a POPC bilayer was calculated for the entire ChR2 homology model across the bilayer (top row) and leaving out TM1 residues modeled as being in a loop (bottom row). The relative rank of differences among the constructs is unchanged under the two conditions. These differences in RHM among constructs are consistent with the pattern of experimentally observed differences in the function of the constructs.

WT (kT)	V259S (kT) (% RHM Decrease)	V269N (kT) (% RHM Decrease)	Q210A (kT) (% RHM Decrease)
12.1	7.7 (36%)	5.8 (52%)	3.1 (74%)
7	3.7 (47%)	3 (56%)	2.3 (67%)

The energy cost of residual hydrophobic mismatch across the bilayer was calculated upon mutagenesis at V269 and Q210 (Table 1). The RHM for Q73 was not calculated because it is located in a modeled loop connecting TMs 1 and 2, which is unresolved in the C1C2 chimera template. Initially, the RHM was calculated for the entire ChR2 homology model, comprising of RHM at residues 210, 269, 270, 69 and 73. As the residues in TM1 are in a modeled loop, RHM was also calculated leaving out the residues in TM1. Both sets of calculations resulted in the same general trends. First, mutation of V269 to serine or asparagine decreased RHM. The decrease in RHM can be directly attributed to the observation that replacement of the hydrophobic residue (V) with polar residues (S or N) eliminates the RHM energy penalty when water penetrates into ChR2 at this position. Similarly, mutation of Q210 to alanine also reduced the RHM energy penalty. Interestingly, the Q210A mutation also lowered RHM at the neighboring V269, as the elimination of the polar residue likely reduced water penetration.

These calculations thus added V269 to the cluster of adjacent polar and hydrophobic residues that are responsible for the large RHM penalties: Q210 (TM6), V269 and L270 (TM7), F69 and Q73 (TM1).

As controls, energy penalty calculations were done for I95L and Q210N, two mutations that should not affect RHM. Indeed, neither of these null mutations changed the energy penalty compared to WT ChR2 (without TM1 contributions, compare the calculated energy penalty value for the WT: 7 kT, to those for I95L: 6.6 kT and Q210N: 6 kT). Differences of <1 kT are within thermal energy fluctuations and cannot be considered biologically significant

3.2. Probing the Effects of Predicted RHM Changes with Experimental Measurements of Function for ChR2 Constructs

In order to directly assess the functional effect of changes in RHM energy penalty on the functional properties of ChR2 constructs, we used two-electrode voltage clamp measurements. A trio of membrane-facing residues located at the cytoplasmic/TM interface (Q73, Q210, V269) were mutated to reduce RHM, in two cases by increasing the hydrophobic surface of the protein (Q73A and Q210A) and in once case decreasing it (V269N). Additionally, the two null mutants (Q210N and I95L) calculated to produce no change in RHM energy cost were also assayed. Functional expression of the mutants was tested in *X. laevis* oocytes under voltage-clamp conditions. The results show reduced photocurrent for the Q73A, Q210A and V269N constructs, while Q210N produced increased photocurrent compared to WT (Figure 3). Q73A had currents <50 nA and was not analyzed further. I95L had no observable photocurrent under our experimental conditions.

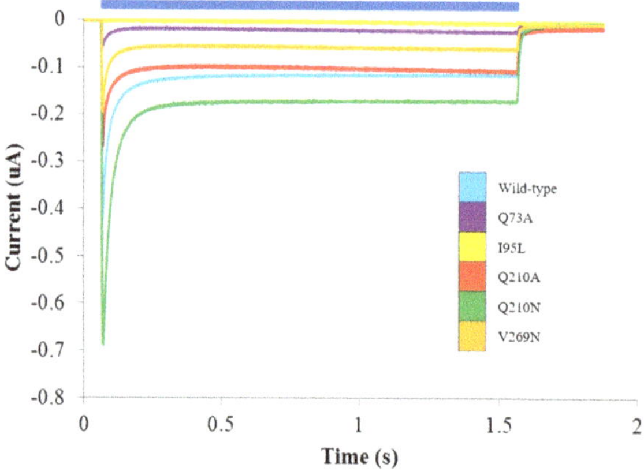

Figure 3. Representative photocurrent traces for RHM ChR2 mutants. Photocurrents were recorded in Na^+ solution at pH 7.5 at $V_m = -100$ mV. Blue bar indicates illumination with blue light. WT—blue; Q73A—purple; I95L—yellow; Q210A—red; Q210N—green; V269N—orange.

The functional effect of RHM on ChR2 selectivity was quantified by measuring reversal potentials (E_{rev}) at I_p and I_{ss} in Na^+ solutions at pH 7.5 and 6.0 (Figure 4A) (see Materials and Methods). This allowed the determination of selectivity between Na^+ and H^+. Additionally, by measuring reversal potentials at the peak and steady-state currents we could monitor the progressive ion selectivity as ChR2 transitions between open states. A positive shift in I_p E_{rev} for V269N was observed at both pH 7.5 and 6.0 compared to WT. The shift was much larger at pH 6.0 than at pH 7.5, where the current is carried mostly by protons (pH 6.0) and not a mixture of Na^+ and H^+ (pH 7.5). In contrast, a hyperpolarizing shift in E_{rev} for I_{ss} was observed for this mutant only at pH 6.0 compared to WT (V269N: 13.2 ± 1.2; WT: 24.6 ± 1.5) (Figure 4B and C; Table 2). This indicated that the permeability of H^+ was increased at the beginning of ion conductance (I_p) but then decreased over time (I_{ss}) when compared to WT. Equally, when changes in E_{rev} were considered, V269N becomes progressively less permeable for H^+ at pH 6.0. Interestingly, the control Q210A mutant had no observable effect on ion permeability. The null mutant did significantly alter the progressive ion selectivity at pH 7.5 and 6.0 but only to a small degree (Table 2). We also determined steady-state to peak current ratios that provide a measure of ChR2 inactivation during prolonged light exposure (Figure 4D). Both Q210A and V269N had significantly increased ratios (Q210A:0.37 ± 0.01; V269N: 0.33 ± 0.01; Vm = −100 mV, pH 7.5) compared to (WT: 0.29 ± 0.01), correlating to less inactivation under continuous illumination. The Q210N mutation had no effect on I_{ss}/I_p ratios.

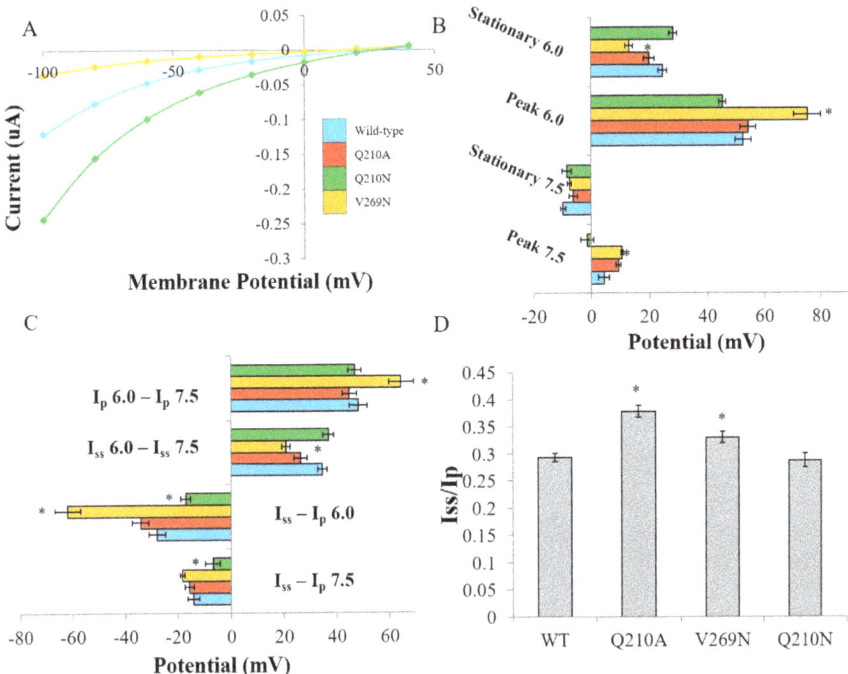

Figure 4. Biophysical characterization of RHM-affecting mutants. (**A**) Representative steady-state current-voltage relationship of select mutants in Na^+ 6.0 solution. (**B**) Absolute reversal potentials determined at I_p and I_{ss} in Na^+ solutions at either pH 6.0 or 7.5. (**C**) Differences in E_{rev} determined as noted. (**D**) Steady-state to peak current ratios, which are a measure of ChR2 inactivation. Values are reported as the average ± SEM (n = 7–15). Statistically significant values are denoted with * ($p < 0.05$). WT—blue; Q210A—red; Q210N—green; V269N—orange.

Apparent rate constants for decay and off kinetics were calculated. Decay kinetics were calculated by fitting ChR2 photocurrent traces from I_p to I_{ss} with a monoexponential or biexponential equation, respectively (Figure 5A). V269N had the largest effect on decay kinetics, resulting in accelerated inactivation. Q210A and the null mutant Q210N had no significant effect on decay kinetics (Figure 5B). A similar trend was observed for off kinetics, with V269N having accelerated rates and both Q210 variants not significantly altering kinetics (Figure 5C).

Table 2. Summary of reversal potentials. The superscript denoted the pH of the solution used. Values are reported at the average ± SEM (n = 7–15). Statistically significant values are denoted with * ($p < 0.05$).

Construct	$I_p^{7.5}$	$I_{ss}^{7.5}$	$I_p^{6.0}$	$I_{ss}^{6.0}$	I_{ss}-$I_p^{7.5}$	I_{ss}-$I_p^{6.0}$	$I_{ss}^{6.0} - I_{ss}^{7.5}$	$I_p^{6.0} - I_p^{7.5}$
WT	4.30 ± 1.9	−9.84 ± 0.8	52.5 ± 2.7	24.6 ± 1.5	−14.1 ± 2.1	−27.8 ± 3.1	34.5 ± 1.7	48.2 ± 3.3
Q210A	9.39 ± 0.8	−6.20 ± 1.5	54.2 ± 2.6	20.1 ± 1.8	−15.6 ± 1.8	−34.1 ± 3.2	26.3 ± 2.4	44.8 ± 2.7
V269N	10.6 ± 0.4	−7.58 ± 0.7	75.0 ± 4.6 *	13.2 ± 1.2 *	−18.2 ± 0.8	−61.8 ± 4.8 *	20.7 ± 1.4 *	64.4 4.7 *
Q210N	−1.55 ± 2.2	−8.5 ± 1.6	45.3 ± 1.2	28.2 ± 1.3	−6.9 ± 2.7 *	−17.0 ± 1.8 *	36.7 ± 2.1	46.9 ± 2.5

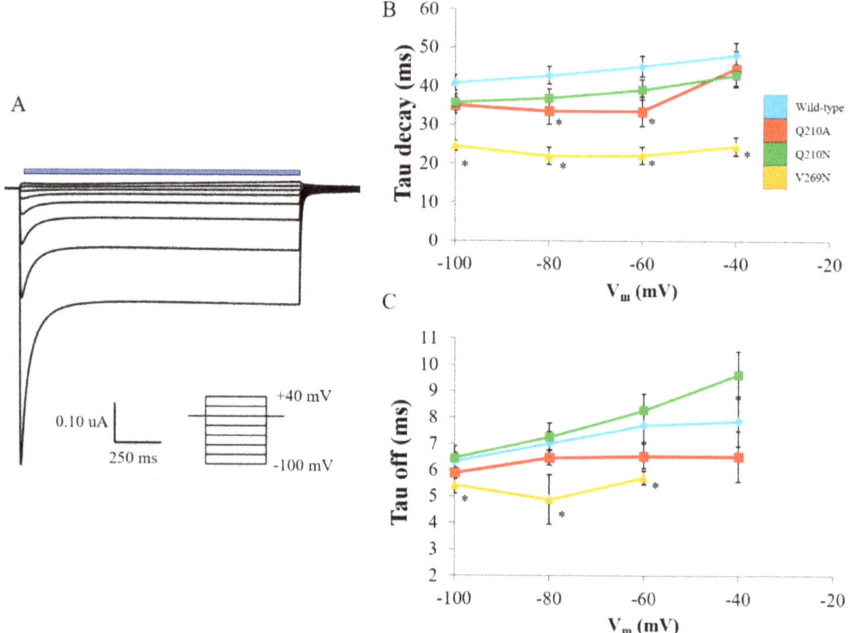

Figure 5. Kinetic properties of RHM-affecting mutants. (**A**) Summary of exponential fitting of apparent rate constants. (**B**) Decay rate comparison of RHM-affecting and null ChR2 mutants. (**C**) Off rate comparison of RHM-affecting and null ChR2 mutants. Error bars represent the SEM (n = 7–15). Statistically significant differences are denoted by * ($p < 0.05$). WT—blue; Q210A—red; Q210N—green; V269N—orange.

Lastly, to determine whether RHM differences explain changes in the modulation of discrete transitions in the ChR2 photocycle, a four-state kinetic model was applied to fit experimental photocurrent traces. Theoretical fits were obtained by optimizing a set of nine parameters describing rates of transition and fundamental properties of the channel (Figure 6) (see Materials and Methods). The null Q210N mutant had the least effect on our parameters compared to WT (Table 3). Only the transition from O1 to O2 was slightly accelerated for this mutant. Both V269N and Q210A saw the largest change from the WT fits, having altered recovery and O2 -> O1 kinetics. Q210A also had a slower O2 to C2 rate while V269N had an accelerated O1 to O2 transition.

Table 3. Summary of parameter optimization for the four-state photocycle model. Values are the average of parameters obtained from each cell ± SEM (n = 3–7). Statistically significant values are denoted with * ($p < 0.05$).

Construct	k1	k2	G_{d1} (ms^{-1})	G_{d2} (ms^{-1})	e_{12} (ms^{-1})	e_{21} (ms^{-1})	Imax	γ	G_r (ms^{-1})	T_{ChR2} (ms)
WT	0.26 ± 0.01	0.081 ± 0.01	0.24 ± 0.03	0.015 ± 0.005	0.038 ± 0.004	0.0057 ± 0.0004	0.57 ± 0.06	0.036 ± 0.007	0.0071 ± 0.001	0.73 ± 0.08
Q210A	0.30 ± 0.001	0.028 ± 0.007	0.16 ± 0.04	0.0055 ± 0.001 *	0.036 ± 0.003	0.010 ± 0.001 *	0.45 ± 0.08	0.050 ± 0.01	0.0093 ± 0.002 *	0.61 ± 0.03
V269N	0.30 ± 0.002	0.037 ± 0.02	0.13 ± 0.05	0.006 ± 0.003	0.061 ± 0.009 *	0.012 ± 0.001 *	0.25 ± 0.05	0.035 ± 0.005	0.030 ± 0.01 *	0.63 ± 0.2
Q210N	0.30 ± 0.03	0.058 ± 0.01	0.34 ± 0.03	0.013 ± 0.005	0.052 ± 0.002 *	0.0067 ± 0.0006	1.7 ± 0.2	0.020 ± 0.004	0.0074 ± 0.001	1.45 ± 0.05

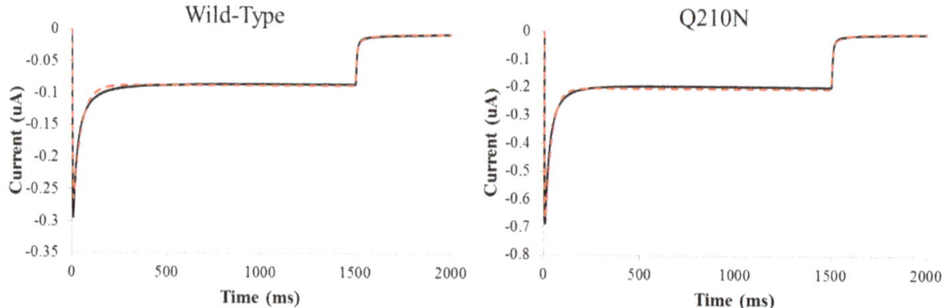

Figure 6. Four-state photocycle theoretical fits. Theoretical fits were obtained by parameter optimization and objective function minimization as described in the Materials and Methods. Solid black line is the experimental average while the dashed red line is the average theoretical fit.

With the four-state model, we explored the population of states at specific time points (Table 4). The time at which each ChR2 construct reached peak current was quantified. Both Q210 variants were comparable to WT (8.0 ms), while V269N was accelerated (5.3 ms). Next, the population of O1 and O2 was evaluated at the time point that corresponded to I_p and also directly before the light was turned off at 1490 ms. Analysis of the two time points revealed similar state populations for all mutants and the WT. The largest variation occurred for O1 at I_p where the RHM mutants Q210A and V269N had higher occupancy while the null Q210N mutant was lower.

Table 4. State population comparison. The population of O1 and O2 were determined at specific time points as described in the text. Values are shown as a fraction of the maximum occupancy, i.e., O1 + O2 + C1 + C2 = 1.

State Population	WT	Q210A	V269N	Q210N
Time to Peak (ms)	8.0	7.7	5.3	7.3
O1 Peak	0.46	0.55	0.52	0.38
O2 Peak	0.09	0.10	0.11	0.09
O1 Stat	0.13	0.19	0.16	0.10
O2 Stat	0.65	0.61	0.67	0.65

4. Discussion

The expression of channelrhodopsin-2 for purposes of optogenetic manipulation has been carried out in a variety of different cell types, including cardiac cells and neuronal cells [5,50]. Importantly, ChR2 undergoes marked conformational changes during photoactivation which are crucial for the gating mechanism and these are likely to affect the interaction with the surrounding membrane environment. Therefore, we undertook here the quantification of an important component of the interaction of ChR2 with a model membrane, employing an approach that combines computational methods that include MD simulations and CTMD analysis [22,51], with related experimental approaches using electrophysiology [32,46] and kinetic modeling [13]. We explored the contribution of the surrounding lipid environment on ChR2 function by quantifying the membrane deformation around the TM core and found that the membrane deformation does not fully eliminate the hydrophobic mismatch between the ChR2 protein and the surrounding membrane, as several residues remain exposed to unfavorable environments. The large energy penalty associated with this RHM was located at the cytoplasm-facing membrane-protein interface located between adjacent TMs 1/2 and 6/7. This is especially interesting in the context of ChR2 gating because of the large movements of TMs 2 and 7 after photoexcitation [16,17,52]. Analysis of MD simulation trajectories showed that the RHM penalty in WT ChR2 is associated with water penetration into the membrane caused by thinning of the bilayer around the polar residues Q210 and Q73. As a result, V269 becomes exposed to a more polar environment. Using experimental electrophysiological approaches, we examined the relation between predicted RHM and measured functional properties to probe the predictive value of the RHM in identifying critical residues in lipid-protein interactions. The simulations were carried out in a generic model of the membrane environment of POPC lipids to enable the evaluation of results in the context of the overwhelming majority of studies of membrane proteins and GPCRs in particular, using this model membrane. These include computational studies by us [19,53–55] and many others [56]. The expectation that membrane-determined effects will result from RHM for the ChR was confirmed by the computational results and the corresponding experimental probing. This opens the way for future studies to interrogate cell-specific environments and their role in the use of the ChR tool, in the way demonstrated in our other studies in which the effects of cholesterol, PIP2 and lipids with varying chain lengths and charge properties [34,35,57,58].

The effects of mutating V269 (in TM7) to asparagine illustrate the relation between a change in the mode of interaction of ChR2 with the membrane and the measured electrophysiological properties. As described, the V269N mutation reduces the hydrophobic mismatch (see Figure 2B) and a hydrophobic residue at this position is highly conserved in channelrhodopsins. The measured reversal potentials for this mutant indicate severely reduced progressive proton selectivity, lower inactivation and accelerated kinetics. Previously, we had observed a similar effect at this position with the V269S mutant as this mutation caused a reduction in channel permeability and pore size [32]. Interestingly, TM7 undergoes reorientation and movement upon photoisomerization [59] and recent MD simulations comparing all-trans retinal and 13-cis retinal bound C1C2 structures revealed outward cytoplasmic bending of TM7 [59]. This is a common movement in microbial-rhodopsins that leads to ion transport [60,61]. Previously, we had attributed the reduction in pore size and function of the V269S mutation to a water-mediated hydrogen bond network that prevents large movements of TM7. However, our experimental results here indicate the importance of lowering RHM at this position and the observed effects on channel permeability for both V269S and V269N in addition to the reduction in pore diameter observed for V269S. This suggests that lowering RHM of ground state ChR2 leads to an unfavorable positioning of TM7 in the open state which restricts the minimum pore diameter. This is further supported by analysis of our kinetic modeling of V269N, which exhibits destabilization between O1 and O2 compared to WT [V269N $R_f = 5.08$; WT $R_f = 6.67$ where R_f is the reaction coefficient (e12/e21)] and reduced maximal current.

We propose that once the pore is formed, V269N becomes exposed to an unfavorable environment or allows for further water penetration exposing proximal hydrophobic residues resulting in the loss of function. This correlates with the conformational movement of TM7 during pore formation and the influx in water prior to ion conductance [15,16,62].

Functionally similar was the effect of Q210 mutation to alanine. Our computational results showed a large reduction in the RHM energy penalty with Q210A. This large reduction is a result of positioning the hydrophobic alanine residue in a favorable environment and also a consequence of the elimination of the Q210 mismatch which decreases the water penetration to the neighboring V269, thereby further lowering the RHM. As functional correlates, we observed a reduction in photocurrent, lower inactivation and slightly accelerated decay kinetics. Although there was no change in selectivity under our experimental conditions, this is unsurprising as TM4 has not been shown to contribute to the permeation pathway. Notably, Q210 is also highly conserved among channelrhodopsins and has a clear structural impact on ChR2 function as suggested by previous serine substitution [32]. The kinetic parameters also indicated a similar destabilization of the open state preference to O2 and significantly reduced quantum efficiency of the C2 to O2 transition as in V269N. To test whether or not our results were a direct consequence of RHM, we created an expected RHM null mutant at this position (Q210N), which we expected to have limited functional impact. Indeed, we found that reversal potentials and kinetics of Q210N are similar to those of WT ChR2. While Q210N did have increased photocurrent, the permeability and kinetics remained similar to WT. Therefore, the increased photocurrent of Q210N is likely not related to RHM. Our modeling also revealed limited impact on photocycle kinetic parameters for this mutant. It remains unclear how decreasing RHM at Q210 affects structural rearrangements of TM4 as little is known about the movement of this TM. However, because mutation of Q210 to alanine also changes the environment at V269 on TM7, it is possible that there is a secondary effect sensed through TM7, which is critical for channel activation.

Our results have documented the clear and quantitatively substantiated effects of residual hydrophobic mismatch at the cytoplasmic interface in ChR2 on channel function. They show that the energy cost of positioning residues in unfavorable environments can indeed be reduced by the exchanges of polar/hydrophobic properties predicted from the computations but that the reduction of the RHM penalty in the closed state has deleterious effects on ChR2 conductance, selectivity and open state stability. This is not always the case, however and the direction of the effect of RHM reduction depends on the functional mechanism, as illustrated by results for the bacterial transporter LeuT, where removal of RHM by the K288A mutant increases transport function nearly 5-fold [63]. The opposite effect of RHM on ChR2 compared to the LeuT case is consonant with the conservation of the RHM-causing residues in the channelrhodopsins, in contrast to the uniqueness of the K288 which is not conserved among LeuT-fold transporters.

Because ChR2 is a conformationally complex channel that is partially regulated by protein-lipid interactions as demonstrated here, it is remarkable that the expression of ChR2 in excitable cells including cardiomyocytes, CA1 pyramidal cells and *Drosophila* motor neurons covers a variety of cell types with diverse lipid composition across species [64–66]. This report, covering what we believe to be the first combined analysis of residual hydrophobic mismatch using computational methods, with prospective experimental validation, shows how exposure of ChR2 residues to unfavorable environments affects channel function. Consequently, the consideration of lipid composition and the specific mutant ChR2 being used for optogenetics merit specific consideration. A more targeted approach can be taken for engineering mutants that will favorably express/function depending on the cell line used.

Author Contributions: Conceptualization, R.R., S.M., H.W. and R.E.D.; Methodology, R.R., S.M., H.W. and R.E.D.; Investigation, R.R. and S.M.; Writing—Original Draft Preparation, R.R., S.M., H.W. and R.E.D.; Funding Acquisition, H.W. and R.E.D.

Funding: This work was supported by the National Institutes of Health grants P01DA012923 and R01MH054137 (Weinstein lab) and DA026434 and the WPI Research Foundation to Robert Dempski.

Acknowledgments: We gratefully acknowledge the allocations of computational resources at (1) the Cofrin Center for Biological Information of the Institute for Computational Biomedicine at Weill Cornell Medical College, (2) the National Energy Research Scientific Computing Center (NERSC, repository m1710) supported by the Office of Science of the U.S. Department of Energy under Contract No. DE-AC02- 05CH11231 (to H.W.) and (3) NSF Teragrid allocations MCB090132 (to R.E.D).

Conflicts of Interest: The authors declare no conflict of interest. The funders had no role in the design of the study; in the collection, analyses or interpretation of data; in the writing of the manuscript or in the decision to publish the results.

References

1. Nagel, G.; Ollig, D.; Fuhrmann, M.; Kateriya, S.; Mustl, A.M.; Bamberg, E.; Hegemann, P. Channelrhodopsin-1: A light-gated proton channel in green algae. *Science* **2002**, *296*, 2395–2398. [CrossRef] [PubMed]
2. Nagel, G.; Szellas, T.; Huhn, W.; Kateriya, S.; Adeishvili, N.; Berthold, P.; Ollig, D.; Hegemann, P.; Bamberg, E. Channelrhodopsin-2, a directly light-gated cation-selective membrane channel. *Proc. Natl. Acad. Sci. USA* **2003**, *100*, 13940–13945. [CrossRef] [PubMed]
3. Richards, R.; Dempski, R.E. From phototaxis to biomedical applications: Investigating the molecular mechanism of channelrhodopsins. In *Electrophysiology of Unconventional Channels and Pores*; Delcour, A.H., Ed.; Springer International Publishing: Cham, Switzerland, 2015; pp. 361–381. [CrossRef]
4. Spudich, J.L.; Sineshchekov, O.A.; Govorunova, E.G. Mechanism divergence in microbial rhodopsins. *Biochim. Biophys. Acta Bioenerg.* **2014**, *1837*, 546–552. [CrossRef] [PubMed]

5. Boyden, E.S.; Zhang, F.; Bamberg, E.; Nagel, G.; Deisseroth, K. Millisecond-timescale, genetically targeted optical control of neural activity. *Nat. Neurosci.* **2005**, *8*, 1263–1268. [CrossRef] [PubMed]
6. Lin, J.Y.; Lin, M.Z.; Steinbach, P.; Tsien, R.Y. Characterization of engineered channelrhodopsin variants with improved properties and kinetics. *Biophys. J.* **2009**, *96*, 1803–1814. [CrossRef] [PubMed]
7. Berndt, A.; Yizhar, O.; Gunaydin, L.A.; Hegemann, P.; Deisseroth, K. Bi-stable neural state switches. *Nat. Neurosci.* **2009**, *12*, 229–234. [CrossRef]
8. Bamann, C.; Kirsch, T.; Nagel, G.; Bamberg, E. Spectral characteristics of the photocycle of channelrhodopsin-2 and its implication for channel function. *J. Mol. Biol.* **2008**, *375*, 686–694. [CrossRef]
9. Berndt, A.; Prigge, M.; Gradmann, D.; Hegemann, P. Two open states with progressive proton selectivities in the branched channelrhodopsin-2 photocycle. *Biophys. J.* **2010**, *98*, 753–761. [CrossRef]
10. Nikolic, K.; Degenaar, P.; Toumazou, C. Modeling and engineering aspects of channelrhodopsin2 system for neural photostimulation. In Proceedings of the Conference of the IEEE Engineering in Medicine and Biology Society, New York, NY, USA, 30 August–3 September 2006; Volume 1, pp. 1626–1629. [CrossRef]
11. Hegemann, P.; Ehlenbeck, S.; Gradmann, D. Multiple photocycles of channelrhodopsin. *Biophys. J.* **2005**, *89*, 3911–3918. [CrossRef]
12. Nikolic, K.; Grossman, N.; Grubb, M.S.; Burrone, J.; Toumazou, C.; Degenaar, P. Photocycles of channelrhodopsin-2. *Photochem. Photobiol.* **2009**, *85*, 400–411. [CrossRef]
13. Richards, R.; Dempski, R.E. Adjacent channelrhodopsin-2 residues within transmembranes 2 and 7 regulate cation selectivity and distribution of the two open states. *J. Biol. Chem.* **2017**, *292*, 7314–7326. [CrossRef] [PubMed]
14. Radu, I.; Bamann, C.; Nack, M.; Nagel, G.; Bamberg, E.; Heberle, J. Conformational changes of channelrhodopsin-2. *J. Am. Chem. Soc.* **2009**, *131*, 7313–7319. [CrossRef] [PubMed]
15. Lorenz-Fonfria, V.A.; Bamann, C.; Resler, T.; Schlesinger, R.; Bamberg, E.; Heberle, J. Temporal evolution of helix hydration in a light-gated ion channel correlates with ion conductance. *Proc. Natl. Acad. Sci. USA* **2015**, *112*, E5796–E5804. [CrossRef] [PubMed]
16. Sattig, T.; Rickert, C.; Bamberg, E.; Steinhoff, H.J.; Bamann, C. Light-induced movement of the transmembrane HelixB in channelrhodopsin-2. *Angew. Chem. (Int. Ed.)* **2013**, *52*, 9705–9708. [CrossRef] [PubMed]
17. Kuhne, J.; Eisenhauer, K.; Ritter, E.; Hegemann, P.; Gerwert, K.; Bartl, F. Early formation of the ion-conducting pore in channelrhodopsin-2. *Angew. Chem. (Int. Ed.)* **2015**, *54*, 4953–4957. [CrossRef]
18. Mondal, S.; Khelashvili, G.; Weinstein, H. Not just an oil slick: How the energetics of protein-membrane interactions impacts the function and organization of transmembrane proteins. *Biophys. J.* **2014**, *106*, 2305–2316. [CrossRef]
19. Mondal, S.; Johnston, J.M.; Wang, H.; Khelashvili, G.; Filizola, M.; Weinstein, H. Membrane driven spatial organization of GPCRs. *Sci. Rep.* **2013**, *3*, 2909. [CrossRef]
20. Chawla, U.; Jiang, Y.; Zheng, W.; Kuang, L.; Perera, S.M.; Pitman, M.C.; Brown, M.F.; Liang, H. A usual g-protein-coupled receptor in unusual membranes. *Angew. Chem. Int. Ed. Engl.* **2016**, *55*, 588–592. [CrossRef]
21. Periole, X. Interplay of G protein-coupled receptors with the membrane: Insights from supra-atomic coarse grain molecular dynamics simulations. *Chem. Rev.* **2017**, *117*, 156–185. [CrossRef]
22. Mondal, S.; Khelashvili, G.; Shi, L.; Weinstein, H. The cost of living in the membrane: A case study of hydrophobic mismatch for the multi-segment protein LeuT. *Chem. Phys. Lipids* **2013**, *169*, 27–38. [CrossRef]
23. Botelho, A.V.; Huber, T.; Sakmar, T.P.; Brown, M.F. Curvature and hydrophobic forces drive oligomerization and modulate activity of rhodopsin in membranes. *Biophys. J.* **2006**, *91*, 4464–4477. [CrossRef] [PubMed]
24. Khelashvili, G.; Weinstein, H. Functional mechanisms of neurotransmitter transporters regulated by lipid-protein interactions of their terminal loops. *Biochim. Biophys. Acta* **2015**, *1848*, 1765–1774. [CrossRef] [PubMed]
25. De Jesus-Perez, J.J.; Cruz-Rangel, S.; Espino-Saldana, A.E.; Martinez-Torres, A.; Qu, Z.; Hartzell, H.C.; Corral-Fernandez, N.E.; Perez-Cornejo, P.; Arreola, J. Phosphatidylinositol 4,5-bisphosphate, cholesterol and fatty acids modulate the calcium-activated chloride channel TMEM16A (ANO1). *Biochim. Biophys. Acta* **2018**, *1863*, 299–312. [CrossRef] [PubMed]
26. Peters, C.J.; Gilchrist, J.M.; Tien, J.; Bethel, N.P.; Qi, L.; Chen, T.; Wang, L.; Jan, Y.N.; Grabe, M.; Jan, L.Y. the sixth transmembrane segment is a major gating component of the tmem16a calcium-activated chloride channel. *Neuron* **2018**, *97*, 1063–1077. [CrossRef] [PubMed]

27. Mondal, S.; Khelashvili, G.; Shan, J.; Andersen, O.S.; Weinstein, H. Quantitative modeling of membrane deformations by multihelical membrane proteins: Application to G-protein coupled receptors. *Biophys. J.* **2011**, *101*, 2092–2101. [CrossRef] [PubMed]
28. Huang, H.W. Deformation free energy of bilayer membrane and its effect on gramicidin channel lifetime. *Biophys. J.* **1986**, *50*, 1061–1070. [CrossRef]
29. Arenkiel, B.R.; Peca, J.; Davison, I.G.; Feliciano, C.; Deisseroth, K.; Augustine, G.J.; Ehlers, M.D.; Feng, G.P. In vivo light-induced activation of neural circuitry in transgenic mice expressing channelrhodopsin-2. *Neuron* **2007**, *54*, 205–218. [CrossRef]
30. Douglass, A.D.; Kraves, S.; Deisseroth, K.; Schier, A.F.; Engert, F. Escape behavior elicited by single, Channelrhodopsin-2-evoked spikes in zebrafish somatosensory neurons. *Curr. Biol.* **2008**, *18*, 1133–1137. [CrossRef]
31. Zhang, F.; Wang, L.-P.; Boyden, E.S.; Deisseroth, K. Channelrhodopsin-2 and optical control of excitable cells. *Nat. Methods* **2006**, *3*, 785–792. [CrossRef]
32. Richards, R.; Dempski, R.E. Re-introduction of transmembrane serine residues reduce the minimum pore diameter of channelrhodopsin-2. *PLoS ONE* **2012**, *7*, e50018. [CrossRef]
33. Shan, J.; Khelashvili, G.; Mondal, S.; Mehler, E.L.; Weinstein, H. Ligand-dependent conformations and dynamics of the serotonin 5-HT(2A) receptor determine its activation and membrane-driven oligomerization properties. *PLoS Comput. Biol.* **2012**, *8*, e1002473. [CrossRef]
34. Khelashvili, G.; Doktorova, M.; Sahai, M.A.; Johner, N.; Shi, L.; Weinstein, H. Computational modeling of the N-terminus of the human dopamine transporter and its interaction with PIP2-containing membranes. *Proteins* **2015**, *83*, 952–969. [CrossRef] [PubMed]
35. Khelashvili, G.; Stanley, N.; Sahai, M.A.; Medina, J.; LeVine, M.V.; Shi, L.; De Fabritiis, G.; Weinstein, H. Spontaneous inward opening of the dopamine transporter is triggered by PIP2-regulated dynamics of the N-terminus. *ACS Chem. Neurosci.* **2015**, *6*, 1825–1837. [CrossRef] [PubMed]
36. Kato, H.E.; Zhang, F.; Yizhar, O.; Ramakrishnan, C.; Nishizawa, T.; Hirata, K.; Ito, J.; Aita, Y.; Tsukazaki, T.; Hayashi, S.; et al. Crystal structure of the channelrhodopsin light-gated cation channel. *Nature* **2012**, *482*, 369–374. [CrossRef] [PubMed]
37. Richards, R.; Dempski, R.E. Cysteine substitution and labeling provide insight into channelrhodopsin-2 ion conductance. *Biochemistry* **2015**, *54*, 5665–5668. [CrossRef] [PubMed]
38. Roberts, E.; Eargle, J.; Wright, D.; Luthey-Schulten, Z. MultiSeq: Unifying sequence and structure data for evolutionary analysis. *BMC Bioinform.* **2006**, *7*, 382. [CrossRef]
39. Volkov, O.; Kovalev, K.; Polovinkin, V.; Borshchevskiy, V.; Bamann, C.; Astashkin, R.; Marin, E.; Popov, A.; Balandin, T.; Willbold, D.; et al. Structural insights into ion conduction by channelrhodopsin 2. *Science* **2017**, *358*, eaan8862. [CrossRef]
40. Yau, W.M.; Wimley, W.C.; Gawrisch, K.; White, S.H. The preference of tryptophan for membrane interfaces. *Biochemistry* **1998**, *37*, 14713–14718. [CrossRef]
41. Ben-Tal, N.; Ben-Shaul, A.; Nicholls, A.; Honig, B. Free-energy determinants of alpha-helix insertion into lipid bilayers. *Biophys. J.* **1996**, *70*, 1803–1812. [CrossRef]
42. Sankararamakrishnan, R.; Weinstein, H. Positioning and stabilization of dynorphin peptides in membrane bilayers: The mechanistic role of aromatic and basic residues revealed from comparative MD simulations. *J. Phys. Chem. B* **2002**, *106*, 209–218. [CrossRef]
43. Gray, T.M.; Matthews, B.W. Intrahelical hydrogen bonding of serine, threonine and cysteine residues within alpha-helices and its relevance to membrane-bound proteins. *J. Mol. Biol.* **1984**, *175*, 75–81. [CrossRef]
44. Eisenhaber, F.; Lijnzaad, P.; Argos, P.; Sander, C.; Scharf, M. The double cubic lattice method—Efficient approaches to numerical-integration of surface-area and volume and to dot surface contouring of molecular assemblies. *J. Comput. Chem.* **1995**, *16*, 273–284. [CrossRef]
45. Nielsen, C.; Goulian, M.; Andersen, O.S. Energetics of inclusion-induced bilayer deformations. *Biophys. J.* **1998**, *74*, 1966–1983. [CrossRef]
46. Gaiko, O.; Dempski, R.E. Transmembrane domain three contributes to the ion conductance pathway of channelrhodopsin-2. *Biophys. J.* **2013**, *104*, 1230–1237. [CrossRef] [PubMed]
47. Talathi, S.S.; Carney, P.R.; Khargonekar, P.P. Control of neural synchrony using channelrhodopsin-2: A computational study. *J. Comput. Biol.* **2011**, *31*, 87–103. [CrossRef] [PubMed]

48. Williams, J.C.; Xu, J.; Lu, Z.; Klimas, A.; Chen, X.; Ambrosi, C.M.; Cohen, I.S.; Entcheva, E. Computational optogenetics: Empirically-derived voltage- and light-sensitive channelrhodopsin-2 model. *PLoS Comput. Biol.* **2013**, *9*, e1003220. [CrossRef]
49. Wang, J.; Hasan, M.T.; Seung, H.S. Laser-evoked synaptic transmission in cultured hippocampal neurons expressing channelrhodopsin-2 delivered by adeno-associated virus. *J. Neurosci. Methods* **2009**, *183*, 165–175. [CrossRef]
50. Bruegmann, T.; Malan, D.; Hesse, M.; Beiert, T.; Fuegemann, C.J.; Fleischmann, B.K.; Sasse, P. Optogenetic control of heart muscle in vitro and in vivo. *Nat. Methods* **2010**, *7*, 897–900. [CrossRef]
51. Khelashvili, G.; Mondal, S.; Andersen, O.S.; Weinstein, H. Cholesterol modulates the membrane effects and spatial organization of membrane-penetrating ligands for G-protein coupled receptors. *J. Phys. Chem. B* **2010**, *114*, 12046–12057. [CrossRef]
52. Krause, N.; Engelhard, C.; Heberle, J.; Schlesinger, R.; Bittl, R. Structural differences between the closed and open states of channelrhodopsin-2 as observed by EPR spectroscopy. *FEBS Lett.* **2013**, *587*, 3309–3313. [CrossRef]
53. Arango-Lievano, M.; Sensoy, O.; Borie, A.; Corbani, M.; Guillon, G.; Sokoloff, P.; Weinstein, H.; Jeanneteau, F. A GIPC1-palmitate switch modulates dopamine drd3 receptor trafficking and signaling. *Mol. Cell. Biol.* **2016**, *36*, 1019–1031. [CrossRef] [PubMed]
54. Sensoy, O.; Weinstein, H. A mechanistic role of Helix 8 in GPCRs: Computational modeling of the dopamine D2 receptor interaction with the GIPC1-PDZ-domain. *Biochim. Biophys. Acta* **2015**, *1848*, 976–983. [CrossRef] [PubMed]
55. Perez-Aguilar, J.M.; Shan, J.; LeVine, M.V.; Khelashvili, G.; Weinstein, H. A functional selectivity mechanism at the serotonin-2A GPCR involves ligand-dependent conformations of intracellular loop 2. *J. Am. Chem. Soc.* **2014**, *136*, 16044–16054. [CrossRef] [PubMed]
56. Kaczor, A.A.; Rutkowska, E.; Bartuzi, D.; Targowska-Duda, K.M.; Matosiuk, D.; Selent, J. Computational methods for studying G protein-coupled receptors (GPCRs). *Methods Cell Biol.* **2016**, *132*, 359–399. [CrossRef] [PubMed]
57. LeVine, M.V.; Khelashvili, G.; Shi, L.; Quick, M.; Javitch, J.A.; Weinstein, H. Role of annular lipids in the functional properties of leucine transporter LeuT proteomicelles. *Biochemistry* **2016**, *55*, 850–859. [CrossRef] [PubMed]
58. Razavi, A.M.; Khelashvili, G.; Weinstein, H. How structural elements evolving from bacterial to human SLC6 transporters enabled new functional properties. *BMC Biol.* **2018**, *16*, 31. [CrossRef] [PubMed]
59. Takemoto, M.; Kato, H.E.; Koyama, M.; Ito, J.; Kamiya, M.; Hayashi, S.; Maturana, A.D.; Deisseroth, K.; Ishitani, R.; Nureki, O. Molecular dynamics of channelrhodopsin at the early stages of channel opening. *PLoS ONE* **2015**, *10*, e0131094. [CrossRef]
60. Nakanishi, T.; Kanada, S.; Murakami, M.; Ihara, K.; Kouyama, T. Large deformation of helix F during the photoreaction cycle of pharaonis halorhodopsin in complex with azide. *Biophys. J.* **2013**, *104*, 377–385. [CrossRef]
61. Radzwill, N.; Gerwert, K.; Steinhoff, H.J. Time-resolved detection of transient movement of helices F and G in doubly spin-labeled bacteriorhodopsin. *Biophys. J.* **2001**, *80*, 2856–2866. [CrossRef]
62. Muller, M.; Bamann, C.; Bamberg, E.; Kuhlbrandt, W. Light-induced helix movements in channelrhodopsin-2. *J. Mol. Biol.* **2015**, *427*, 341–349. [CrossRef]
63. Piscitelli, C.L.; Krishnamurthy, H.; Gouaux, E. Neurotransmitter/sodium symporter orthologue LeuT has a single high–affinity substrate site. *Nature* **2010**, *468*, 1129–1132. [CrossRef] [PubMed]
64. Hamazaki, K.; Kim, H.-Y. Differential modification of the phospholipid profile by transient ischemia in rat hippocampal CA1 and CA3 regions. *ProstaglandinsLeukot. Essent. Fat. Acids* **2013**, *88*, 299–306. [CrossRef] [PubMed]
65. Jones, H.E.; Harwood, J.L.; Bowen, I.D.; Griffiths, G. Lipid composition of subcellular membranes from larvae and prepupae of *Drosophila melanogaster*. *Lipids* **1992**, *27*, 984–987. [CrossRef] [PubMed]
66. Norton, W.T.; Abe, T.; Poduslo, S.E.; DeVries, G.H. The lipid composition of isolated brain cells and axons. *J. Neurosci. Res.* **1975**, *1*, 57–75. [CrossRef] [PubMed]

Article

Atomistic Insight into the Role of Threonine 127 in the Functional Mechanism of Channelrhodopsin-2

David Ehrenberg [1,†], Nils Krause [2,†], Mattia Saita [1], Christian Bamann [3], Rajiv K. Kar [4], Kirsten Hoffmann [2], Dorothea Heinrich [2], Igor Schapiro [4], Joachim Heberle [1,*] and Ramona Schlesinger [2,*]

1 Experimental Molecular Biophysics, Department of Physics, Freie Universität Berlin, Arnimallee 14, 14195 Berlin, Germany; david.ehrenberg@fu-berlin.de (D.E.); mattiasaita@zedat.fu-berlin.de (M.S.)
2 Genetic Biophysics, Department of Physics, Freie Universität Berlin, Arnimallee 14, 14195 Berlin, Germany; NilsKrause@web.de (N.K.); hoffmank@zedat.fu-berlin.de (K.H.); dorothea.heinrich@fu-berlin.de (D.H.)
3 Max-Planck-Institute of Biophysics, Max-von-Laue-Straße 3, 60438 Frankfurt am Main, Germany; christian.bamann@dfg.de
4 Fritz Haber Center for Molecular Dynamics Research, Institute of Chemistry, Hebrew University of Jerusalem, Jerusalem 9190401, Israel; rajiv.kar@mail.huji.ac.il (R.K.K.); igor.schapiro@mail.huji.ac.il (I.S.)
* Correspondence: joachim.heberle@fu-berlin.de (J.H.); r.schlesinger@fu-berlin.de (R.S.); Tel.: +49-30-838-56161 (J.H.); +49-30-838-56249 (R.S.)
† These authors contributed equally.

Received: 28 September 2019; Accepted: 12 November 2019; Published: 15 November 2019

Abstract: Channelrhodopsins (ChRs) belong to the unique class of light-gated ion channels. The structure of channelrhodopsin-2 from *Chlamydomonas reinhardtii* (*Cr*ChR2) has been resolved, but the mechanistic link between light-induced isomerization of the chromophore retinal and channel gating remains elusive. Replacements of residues C128 and D156 (DC gate) resulted in drastic effects in channel closure. T127 is localized close to the retinal Schiff base and links the DC gate to the Schiff base. The homologous residue in bacteriorhodopsin (T89) has been shown to be crucial for the visible absorption maximum and dark–light adaptation, suggesting an interaction with the retinylidene chromophore, but the replacement had little effect on photocycle kinetics and proton pumping activity. Here, we show that the T127A and T127S variants of *Cr*ChR2 leave the visible absorption maximum unaffected. We inferred from hybrid quantum mechanics/molecular mechanics (QM/MM) calculations and resonance Raman spectroscopy that the hydroxylic side chain of T127 is hydrogen-bonded to E123 and the latter is hydrogen-bonded to the retinal Schiff base. The C=N–H vibration of the Schiff base in the T127A variant was 1674 cm^{-1}, the highest among all rhodopsins reported to date. We also found heterogeneity in the Schiff base ground state vibrational properties due to different rotamer conformations of E123. The photoreaction of T127A is characterized by a long-lived P_2^{380} state during which the Schiff base is deprotonated. The conservative replacement of T127S hardly affected the photocycle kinetics. Thus, we inferred that the hydroxyl group at position 127 is part of the proton transfer pathway from D156 to the Schiff base during rise of the P_3^{530} intermediate. This finding provides molecular reasons for the evolutionary conservation of the chemically homologous residues threonine, serine, and cysteine at this position in all channelrhodopsins known so far.

Keywords: channelrhodopsin; resonance Raman; flash photolysis; hybrid QM/MM simulation; electrophysiology

---

1. Introduction

Channelrhodopsins (ChRs) are members of the group of microbial rhodopsins that are light-gated cation channels. They were originally found in the eyespot of the algae *Chlamydomonas reinhardtii* (*Cr*), where they serve to identify optimal light conditions during phototactic movement [1,2]. Typically for rhodopsins, ChRs consist of seven α-helices spanning the membrane and a retinylidene chromophore, which is bound via a Schiff base to a lysine in the core of the protein.

Within the new research field of optogenetics, the coding sequences for light-responsive proteins together with regulating promoters can be introduced into complex organisms and expressed in a tissue-specific way, where cell processes can be triggered. Channelrhodopsin-2 (*Cr*ChR2) in neurons can depolarize the nerve cell when illuminated by light of the corresponding wavelength. Although this protein is already frequently used as optogenetic tool, aimed at medical approaches like restoring vision and hearing, the mechanism of action is still not completely understood. Comparison to related microbial rhodopsins where individual positions in the protein sequences are highly conserved could guide our approach towards mechanistically important amino acid residues which have demonstrated effects on photocycle kinetics, activity, and/or structural stability upon mutation.

Decades ago, the Khorana group investigated the role of threonines in the proton pump bacteriorhodopsin (bR) from *Halobacterium salinarum*. In that study, the T89V variant exhibited a blueshift of 146 nm of the visible absorption spectrum without impairing proton pumping [3]. It was concluded from FTIR and time-resolved absorption spectroscopy in the visible range that T89 exerts steric constraints during isomerization of the chromophore and is part of a hydrogen-bonded network including a water molecule and the residues Y185 and D212 [4]. Years later, the same group demonstrated that the hydroxyl group of T89 interacts with the protonated Schiff base in light-adapted bR and influences proton transfer to D85 during the photocycle [5]. However, from X-ray structures [6,7], it became apparent that the OH group of T89 forms a hydrogen bond to an oxygen of the side chain of D85 rather than to the Schiff base. It was concluded from polarized FTIR spectroscopy that the isomerization of the chromophore after light excitation leads to a shortening of the hydrogen bond between T89 and D85, indicating an interconnection to the Schiff base [8]. This tight complex persists from the K intermediate until the M intermediate [9].

The hydroxylic amino acid side chain of T89 of bR is conserved in many channelrhodopsins. The corresponding residue T127 of *Cr*ChR2 is located in immediate vicinity to the retinal Schiff base (Figure 1), with a distance of 3.7 Å of the threonine oxygen to the Schiff base nitrogen in the crystal structure. Other channelrhodopsins carry a serine [10] or a cysteine at this position. The latter appears in *Ca*ChR1 (from *Chlamydomonas augustae*) and *Cy*ChR1 (from *Chlamydomonas yellowstonensis)* [11]. Even anion-conducting channelrhodopsins, which do not conserve E123, have a threonine or a cysteine at this position [12].

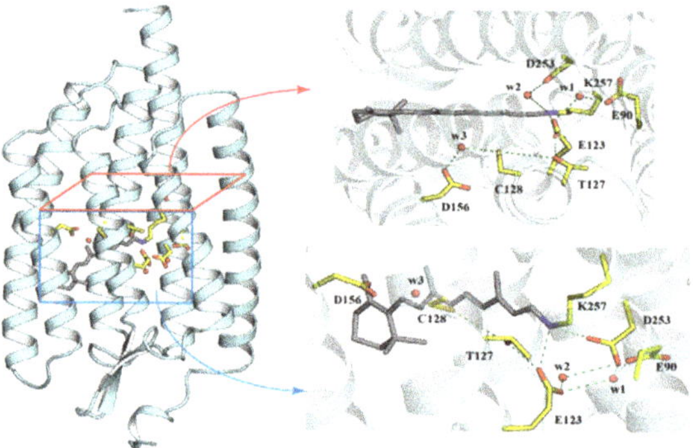

Figure 1. X-ray crystallographic structure of *Chlamydomonas reinhardtii* channelrhodopsin-2 (*Cr*ChR2) (PDB: 6EID [10]). The protonated Schiff base of retinal and the side chains of key residues are shown as grey and yellow sticks, respectively. Water molecules are shown as red spheres. Dashed lines indicate hydrogen-bonding interaction.

In *Cr*ChR2, the residue T127 is connected via a hydrogen bond to E123, the homologous residue of D85 in bR [10]. The protonation state of E123 is not clear, but nearby ionized D253 serves as the acceptor of the Schiff base proton during the rise of the P_2^{390} intermediate [13]. D156 has been shown to interact via hydrogen bonding with C128, denoted as the DC gate [14]. The X-ray structure as well as theoretical calculations confirmed that the terminal groups of both amino acids are interconnected via a water molecule [10,15]. D156 is protonated in the dark state and deprotonates during the P_2^{390} to P_3^{520} transition concomitantly to the reprotonation of the Schiff base. Thus, D156 is the proton donor of the Schiff base in *Cr*ChR2 [13]. The distance from the DC gate to the Schiff base is too large for direct proton transfer. It was shown in Reference [10] that the nearby residue T127 may facilitate the reprotonation process. To evaluate this hypothesis, we set out to investigate variants in which T127 was replaced by alanine or serine.

Channel activity of the T127 variants was examined by electrophysiology and exhibited reduced conductance in the T127 variants. Molecular spectroscopy (UV/Vis, FTIR, Raman) was applied to scrutinize the role of T127 in the molecular mechanism of *Cr*ChR2. We report here the exceptionally high frequency of the Schiff base vibration in the T127A variant, indicating strong interaction between the protonated Schiff base and the carboxylic side chain of E123 in this variant. QM/MM calculations supported and extended the spectroscopic results with detailed atomistic descriptions of the hydrogen-bonded network surrounding the retinal Schiff base. We believe that the characterization of the threonine variants at the molecular level is an important step towards the understanding of the link between the protein's spectroscopic properties and its function as a light-activated cation channel.

2. Materials and Methods

2.1. Site-Directed Mutagenesis, Cloning, Expression and Purification of CrChR2

The experiments were executed with recombinant *Cr*ChR2, which consisted only of the membranous part containing amino acid residues 1–307 of channelrhodopsin-2 of *Chlamydomonas reinhardtii* (UniProt: Q8RUT8). Due to insertion of the corresponding coding region with sequences for a C-terminal 10xHis tag and a linker (aa AS) behind the alpha factor signal sequence of the vector pPIC 9K into the *Eco*RI/*Not*I sites, the expressed protein led to a N-terminal extension of aa YVEFH and a C-terminal extension of aa ASHHHHHHHHHH. Based on this construct, which is referred in the following as *Cr*ChR2-WT or simply WT, T127 was substituted by serine (T127S) or alanine residues (T127A). The substitutions were introduced with oligonucleotide-directed mutagenesis using PCR and verified by sequencing of the *Cr*ChR2 coding sequence. The *Cr*ChR2 variants were expressed and purified as described previously [16].

2.2. Molecular Spectroscopy

Time-resolved UV/Vis experiments were performed with a commercial flash photolysis unit (LKS80, Applied Photophysics, Leatherhead, Surrey, UK), as described previously [17]. Briefly, a 10 ns laser pulse (Nd:YAG, Quanta-Ray, Spectra-Physics) tuned to 450 nm by an optical parametric oscillator (OPTA) was used to induce the photocycle. The energy density per pulse was set to 3 mJ/cm^2. Five time traces were averaged at each wavelength with a repetition frequency of 0.33 Hz.

Samples used for FTIR spectroscopy were concentrated to ~4 mg/mL *Cr*ChR2 in an aqueous solution of 20 mM Hepes and 0.2% DM (n-decyl-β-D-maltopyranoside) at pH 7.4. For the FTIR experiments, approximately 8 µl of *Cr*ChR2 was dried on top of a BaF_2 window. The protein film was rehydrated with the saturated vapor phase of a glycerol/water mixture (2/8 w/w) [18] and placed into the FTIR spectrometer (Vertex 80v, Bruker, Rheinstetten, Germany). Time-resolved rapid-scan FTIR spectroscopy was employed to resolve intermediates with a time resolution of about 10 ms.

Resonance Raman experiments were performed essentially as described in Reference [19]. Briefly, 5 µL of concentrated sample (5–10 mg/mL) was dried and rehydrated on a quartz crucible and subsequently cooled to 80 K using a N_2 cryostat (Linkham). Rehydration occurred prior cooling via vapor diffusion of 3 µl of either H_2O/glycerol or D_2O/glycerol mixtures (8/2 w/w) placed in the vicinity of the sample. The emission line at 457 nm of a diode-pumped solid-state (DPSS) laser (Changchun New Industries Optoelectronics Technology Co., Ltd., China) was used to induce Raman scattering.

2.3. Electrophysiology

Light-induced currents were recorded from oocytes in two-electrode voltage-clamp (TEVC) experiments after expression of the WT and the T127A or T127S variants for at least 3 days. To this end, in-vitro-synthesized RNA coding for a fusion protein of chop2 (residues 1 to 315) and egfp was injected into oocytes and incubated in oocyte Ringer (ORI, 90 mM NaCl, 2 mM KCl, 2 mM $CaCl_2$, 1 mM $MgCl_2$, 5 mM Hepes, pH 7.4) solution supplemented with 5 µM all-trans retinal. Oocytes were illuminated with light from a 75 W XBO lamp long-pass filtered at 420 nm in ORI. Currents of the T127 variants were normalized to the WT current amplitudes at −60 mV from oocytes recorded at the same day.

Time-resolved single turnover currents of the T127 variants were also recorded from HEK293 cells expressing the same fusion protein as in the oocyte recordings. Cells were illuminated with a single flash (20 ns) from an Excimer-laser pumped dye laser (coumarin 2, 450 nm). Traces were sampled at 50 kHz and filtered at 10 kHz. The displayed traces are averages of eight signals.

2.4. QM/MM Calculations

The structural model for hybrid quantum mechanics/molecular mechanics (QM/MM) calculations was based on the crystal structure of *Cr*ChR2 (PDB entry: 6EID [10]) with an all-trans conformation of the retinal protonated Schiff base ($RSBH^+$). Titratable residues were modeled corresponding to the experimental pH and considering their local environment. Based on spectroscopic studies, the residue E90 was modeled in the neutral form and the counterions of $RSBH^+$ (E123 and D253) were taken as negatively charged. The variants of *Cr*ChR2, T127A and T127S, were prepared with PyMol [20].

In order to account for the effect of T127 and its variants, we used a large QM region that comprised the $RSBH^+$ side chains of E123, T127, A127, S127, C128, D253, K257, and two water molecules in the close vicinity of $RSBH^+$. The hydrogen link atom was placed at the QM/MM boundary between the $C\delta$ and $C\varepsilon$ atoms of K257, and between the $C\alpha$ and side chain atom of remaining residues. The QM region was treated at the B3LYP/cc-pVDZ level of theory [21]. Corrections for the dispersion effect were included with D3/B-J damping variant [22]. All the remaining atoms were described at the MM level using the AMBER ff14SB force field [23]. The vibrational frequencies were calculated at the same level of theory. A scaling factor of 0.97 was used [24]. However, the QM region in frequency calculations did not include the water molecules. Furthermore, the QM/MM model was used to calculate the excitation energies using the simplified TD-DFT approach developed by Grimme and coworkers [25]. All QM/MM calculations were carried out using ChemShell interfaced with the quantum chemistry program Orca with DL_POLY module for the force field [26].

3. Results

In the X-ray crystal structure of *Cr*ChR2, the distance between heavy atom pairs $N(RSBH^+)\cdots O(E123)$ and $N(RSBH^+)\cdots O(D253)$ is 2.81 and 3.24 Å, respectively. The QM/MM geometry optimization leads to slight adjustments in the retinal binding pocket, which modifies the corresponding distances to 2.82 and 3.48 Å. In WT, T127 was found to be hydrogen-bonded with E123 with an $OH(T127)\cdots O(E123)$ distance of 1.70 Å (Figure 2A). Upon T127S exchange, the hydrogen-bonding network was retained with a slightly reduced distance of 1.67 Å between $OH(T127S)\cdots O(E123)$ and an increased distance of 3.53 Å between $N(RSBH^+)\ldots O(D253)$ (Figure 2B). The difference between serine and threonine residues is the absence of the methyl group at the C_β position in the former. Lack of this bulky group in serine might cause a reduced steric effect responsible for its approach to E123. In contrast, the T127A variant was missing a proton-donating group (Figure 2C); therefore, E123 had only one hydrogen-bonding partner, namely the Schiff base. The distance between $N(RSBH^+)\cdots O(E123)$ remained the same (2.81 Å), but the N–H bond length was elongated by 0.016 Å (Table 1) because the proton is pulled more strongly towards E123 as a consequence of its altered hydrogen bonding.

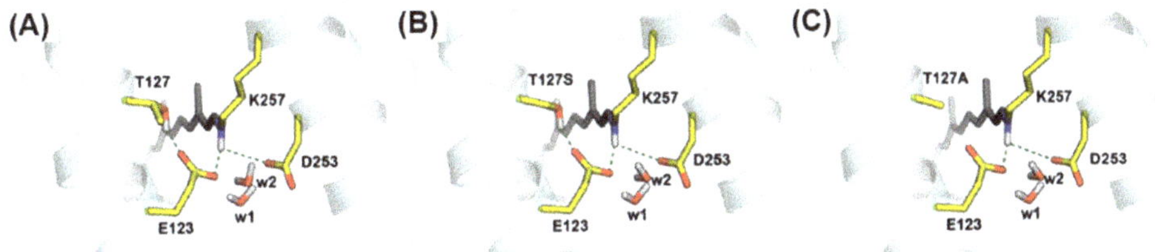

Figure 2. Optimized geometry of (**A**) *Cr*ChR2-WT, (**B**) T127S, and (**C**) T127A variants. The retinal Schiff base and side chains of the counterion complex are shown as grey and yellow sticks, respectively. Dotted lines correspond to the hydrogen-bonding network.

Table 1. Bond length in selected RSBH$^+$ atom pairs, based on the hybrid quantum mechanics/molecular mechanics (QM/MM)-optimized geometries.

RSBH$^+$ Atom Pairs (Distances in Å)	WT	T127S	T127A
C=N	1.310	1.308	1.309
N–H	1.076	1.072	1.092

In *Cr*ChR2-WT, T127 was also found to be involved in a triad between O(E123), OH(T127), and S(C128). The corresponding distances in the optimized geometry were found to be 4.91 Å for O(E123) ··· S(C128) and 4.39 Å for OH(T127) ··· S(C128). In the T127S variant, the distance between O(E123) ··· S(C128) was slightly reduced to 4.88 Å, whereas the distance between OH(T127S) ··· S(C128) was increased to 4.47 Å. This indicated that the T127S variant led to a tighter network with the DC gate and the SB counterions. However, the T127A exchange perturbed the triad and decoupled the DC gate from the counterions, as the distance between O(E123) ··· S(C128) was increased to 4.95 Å.

3.1. *Spectroscopic Properties of the Dark State of the T127 Variants of CrChR2*

As T127 is part of the retinal binding pocket, we first examined the potential influence of the amino acid exchange on the electronic properties of the retinal. It was evident from comparison of the UV/Vis spectra (Figure 3, red and blue spectra) that both T–A and T–S amino acid exchanges shifted the visible absorption spectrum only slightly to the blue as compared to WT (black spectrum in Figure 3). The second derivative of the absorption spectra (bottom panel) resolved a blue shift of about 2 nm in the T127A variant as compared to the WT (similar to T127S, not shown). The vibronic fine structure of the electronic absorption spectrum was retained in the T127A and the T127S variants. The latter observation was different to the C128T variant, where a 20 nm redshift of the retinal absorption was accompanied by a loss in vibronic fine structure [27].

We also computed the excitation energies for the WT protein and the two variants. The results displayed in Table 2 showed virtually no alteration of the excitation energies, which agreed well with the experimental results (Figure 3). These results present an additional validation of our QM/MM model.

Table 2. Computed excitation energies at the sTD-DFT CAM-B3LYP/cc-pVDZ level of theory.

Model	WT	T127S	T127A
Excitation energy nm (eV)	389 (3.18)	392 (3.16)	388 (3.19)

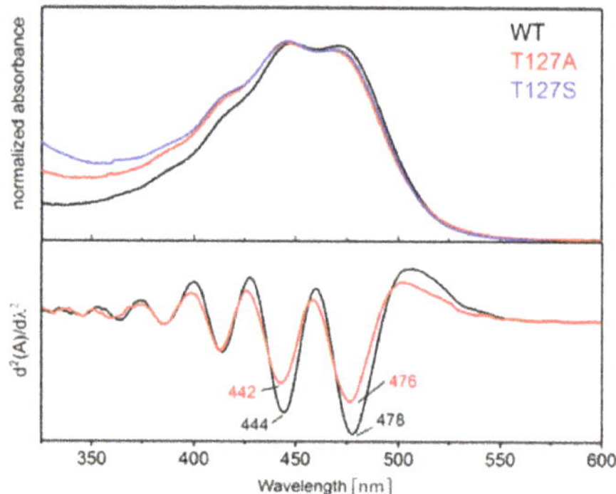

Figure 3. UV/Vis absorption spectra for the *Cr*ChR2-WT, T127A, and T127S variants. The second derivatives of the absorption spectra are displayed in the bottom panel; the second derivative of the T127S spectrum has been omitted for clarity as it overlaps with the T127A trace.

As the wavelength of the electronic absorption in the visible correlated with the C=C stretching vibration [28], the strongest band in the resonance Raman spectrum of the T127A variant (Figure 4, green trace) was 1557 cm^{-1}, slightly blue-shifted by 6 cm^{-1} with respect to the WT (Figure 5A, [29]). All of the other Raman bands were essentially the same as in the WT, except for the C=N–H vibration of the retinal Schiff base (Figure 5B). The frequency of the latter was observed at 1674 cm^{-1} for T127A, which is, to our knowledge, the highest frequency so far observed for any rhodopsin. It even outperformed the frequency of the Schiff base vibration in rhodopsin (1660 cm^{-1}, [30]). The C=N–H vibrational band was downshifted by 42 cm^{-1} upon H/D exchange, which was the largest isotope effect reported (Figure 4). In comparison, the C=N–H stretch in WT resonated with a frequency of 1657 cm^{-1} and a $\Delta\upsilon_{H/D}$ = 28 cm^{-1} was reported [29]. As the hydrogen bond between T127 and E123 was removed in the T127A variant (Figure 2C), the carboxylic side chain of E123 was able to form a stronger salt bridge with the retinal Schiff base, resulting in a 0.016 Å elongated N–H bond (Table 1).

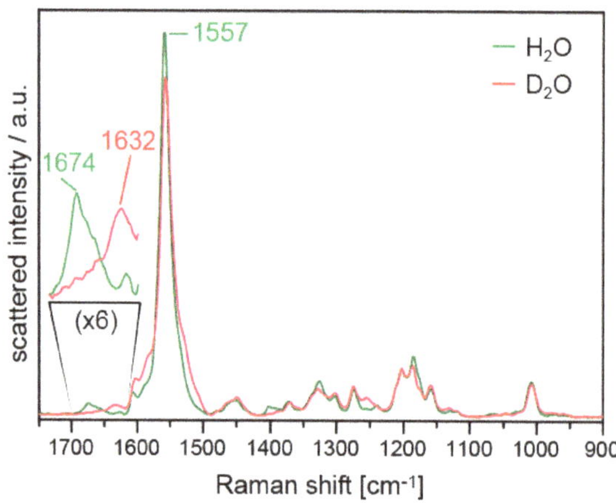

Figure 4. Resonance Raman spectra of the T127A variant in H_2O (green trace) and in D_2O (red trace). The indicated vibrational bands are discussed in the text.

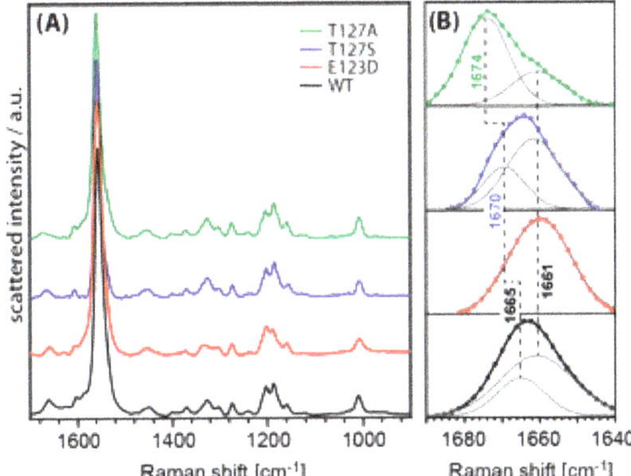

Figure 5. (**A**) Resonance Raman spectra of *Cr*ChR2-WT and the variants E123D, T127S, and T127A in H_2O. (**B**) Zoom-in of the frequency range of the Schiff base vibration after a three-point moving average smooth. Blue circles are the data points. Dashed grey lines are Gaussians fitted to the data points and continuous lines are sums of the fitted Gaussians.

Inspection of the vibrational band at 1674 cm^{-1} revealed a strong asymmetry in the shape, with a pronounced shoulder at lower wavenumbers (green trace in Figure 5B). Consequently, the band was fitted by two Gaussians with frequency maxima at 1674 cm^{-1} and 1661 cm^{-1}. Such heterogeneity in the retinal binding pocket has been suggested from molecular dynamics simulations on WT [31] to be a result of two rotamers of the side chain of E123. In one rotamer configuration, E123 was in direct hydrogen-bonding interaction with the protonated Schiff base, competing with a water molecule and D253. The other rotamer configuration positioned the carboxylic group of E123 in a remote position from the Schiff base. A water molecule or the carboxylic group of D253 can then form a hydrogen bond with the Schiff base proton. The resulting effect on the strength of the N–H bond of the Schiff base was acutely determined by Raman spectroscopy.

The vibrational band of the Schiff base in WT has so far been treated as a single vibration [32]. In light of our results for the T127A variant, we revisited the Raman spectrum of the WT (Figure 5, black trace). The slight asymmetry in band shape was again considered by fitting two Gaussians, resulting in maxima at 1665 cm^{-1} and 1661 cm^{-1}.

From these results, we assigned the lower frequency of the C=N–H vibration at 1661 cm^{-1} to the Schiff base being hydrogen-bonded to a water molecule and/or directly to D253. In this scenario, the substitution of T127 by A did not affect the frequency of the Schiff base vibration, as this residue did not interact directly with D253. This explained the low-frequency shoulder of the C=N–H band at 1661 cm^{-1} which did not shift after amino acid exchange. The high-frequency band at 1674 cm^{-1} in T127A corresponded to the population in which E123 was hydrogen bonded to the Schiff base and, with respect to the WT, lacked the hydrogen bond to the threonine (see Figure 2, right panel).

We tested this hypothesis by recording resonance Raman spectra for two other *Cr*ChR2 variants: T127S, where the hydroxyl group of the threonine side chain is conserved; and E123D, where the carboxylate side chain is shorter. In agreement with the WT and the T127A variant, all spectral features of the retinal chromophore were retained in the spectra of the T127S (blue traces in Figure 5) and E123D variants (red traces in Figure 5) with only minor differences. Analysis of the Schiff base vibration (Figure 5b) showed that the C=N–H band in E123D could be fitted sufficiently well with one Gaussian, whereas for T127S, two Gaussians were needed to fit the band. The band with a maximum at 1661 cm^{-1} was present in both variants, supporting the hypothesis that this Schiff base vibration did not involve interaction with E123.

In T127S, where the exchange retained the hydroxyl group that is hydrogen-bonded to E123 (see Figure 2), the high-frequency shoulder of the peak was only slightly blue-shifted by 5 cm^{-1}, showing strong similarities to the C=N–H vibrational band of the WT. In E123D, however, the high frequency band was not present at all as a consequence of the shorter amino acid side chain length that prevented formation of a hydrogen bond with the Schiff base.

We focused our QM/MM calculations on the structure with the rotamer configuration of E123 in the upward orientation to be able to accept a hydrogen bond from the Schiff base proton. This configuration was in fact the one in which the C=N–H vibration exhibited the strongest effect in our T127 variants. Although the spectroscopic data showed that the hydrogen bond between the Schiff base and E123 was strongly strengthened upon replacement of T127, the distance between the Schiff base nitrogen and E123 was unchanged in the optimized QM/MM structure of the T127 variants. Thus, we went on to analyze the effect of the amino acid exchange on the length of the C=N and the N–H bonds of the Schiff base in our simulations (Table 1).

The length of the C=N bond was not significantly affected in T127S and T127A as compared to the WT. However, the hydrogen-bonding network in the vicinity of $RSBH^+$ was found to have a notable effect, with the presence or absence of hydroxylic group in the sidechain at position 127. The difference in N–H bond length in WT and T127S was negligible (~0.004 Å), which was attributed to the hydrogen-bonding network facilitated by the hydroxylic side chain. In contrary, the absence of such a network in T127A increased the N–H bond length by 0.016 Å. The longer N–H bond length in T127A was in line with our experimental findings that this variant accelerated deprotonation of $RSBH^+$ (see Section 3.3) and it also supported the high frequency of the C=N–H vibration of 1674 cm^{-1}. QM/MM simulations yielded a frequency difference of 19 cm^{-1} between WT and the T127A variant (Table 3), which reproduced the frequency upshift upon removal of the hydroxylic group of threonine but overestimated the experimental difference of 9 cm^{-1} (Figure 5B). The exceptionally high frequency of the Schiff base C=N–H vibration in the T127A variant was, therefore, be assigned to the missing hydrogen-bond between A127 and E123 and the increase in distance between $N(RSBH^+) \cdots O(E123)$.

Table 3. Calculated vibrational frequencies of the C=N–H vibration using the QM/MM method.

Model	WT	T127S	T127A
Frequencies (cm^{-1})	1683 cm^{-1}	1683 cm^{-1}	1702 cm^{-1}

3.2. *Channel Conductance of the T127 Variants*

The role of T127 in the functional mechanism of the ion channel was scrutinized via electrophysiological experiments on WT and its T127 variants expressed in *Xenopus* oocytes. The T127A and the T127S variants showed a strong decrease in the current amplitude (for T127A: Figure 6A, red traces). On average, the variants exhibited only 9 ± 1% (n = 29 for T127A and n = 18 for T127S) of the WT currents. From the residual current, we determined the kinetics of the channel closing upon switching off the light (Figure 6B). It was slower than the WT (τ = 12 ms), with τ = 22 ± 3 ms (n = 9) for the T127A variant and with τ = 23 ± 3 ms (n = 9) for the T127S variant.

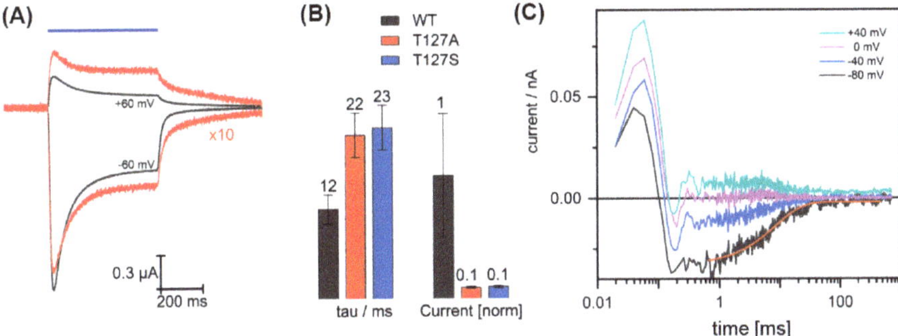

Figure 6. Two-electrode voltage-clamp (TEVC) recordings of *Cr*ChR2-WT, T127A, and T127S variants expressed in *Xenopus* oocytes. (**A**) Light-induced currents at −60 mV and +60 mV holding potential. The traces for the T127A variant (red) were scaled by a factor of 10 to compare to the WT (black). (**B**) Histogram of the current amplitudes and closing kinetics. Currents were normalized to the WT at −60 mV. Bars represent the mean and the s.e.m. (n = 14 for WT, n = 29 for T127A variant, and n = 18 for T127S variant). (**C**) Time-resolved currents from T127A in HEK293 cells. Raw data of an average of eight traces at −80 mV to +40 mV in steps of 40 mV. The red curve shows a single exponential fit of the current decay with a time constant of 10 ms.

In time-resolved experiments on the T127A variant expressed in HEK293 cells, only very small current amplitudes were recorded. Signals from eight kinetic traces were averaged (Figure 6C) to be able to determine the channel closing kinetics. Similarly to the E123T variant [33] we observed a very fast outward current independent of the holding potential, which monitored a vectorial charge transport within the first 200 μs. The recovery kinetics after pulsed excitation exhibited single exponential behavior with a time constant of 10 ms, which was similar to the WT and consistent with our results from the TEVC recordings.

3.3. Influence of T127 on the Photocycle Kinetics

As the channel conductance was drastically reduced by the T127 replacement but the absorption spectrum of dark-state *Cr*ChR2 was hardly affected, we performed time-resolved UV/Vis spectroscopy (Figure 7). Excitation by a nanosecond laser pulse led to the depletion of the dark state of *Cr*ChR2, which was reflected by a loss of absorption at 470 nm (blue traces in Figure 7). The P_1^{500} state rose at times beyond the resolution of our experiment, but the decay was resolved at 530 nm (green traces). The rise of the blue-shifted P_2^{380} state with a deprotonated retinal Schiff base was recorded at 380 nm (red traces). Rise and decay of the succeeding P_3^{530} intermediate were observed at 530 nm and the latter correlated with channel closure in *Cr*ChR2 [34]. The kinetics of the desensitized state P_4^{480} were also observed at 530 nm.

The P_2^{380} state was formed faster in the T127A variant (Figure 7, top panel) and decayed slower than in the WT (Figure 7, bottom panel), i.e., the Schiff base deprotonated at an earlier stage and was reprotonated later. The kinetics of P_2^{380} rise in T127A had a half-life of $t_{1/2}$ = 0.7 μs, one order of magnitude faster than the rise of the P_2^{380} state in WT. The decay of this blue-shifted intermediate in T127A, with a $t_{1/2}$ = 9 ms, was more than four times slower than in WT. The decay kinetics of the P_2^{380} state were altered at the expense of the P_3^{530} intermediate, which rose later and was significantly reduced in transient amplitude. This observation tallied the reduced channel conductance based on the correlation of the lifetimes of the open state and the P_3^{530} state. The decay of the non-conductive P_4^{480} state was not influenced by the T127A exchange. The less invasive replacement of threonine by serine in the T127S variant left the photocycle kinetics of *Cr*ChR2 unchanged (Figure 7, middle panel). Thus, we inferred that the hydroxyl group at position 127 was essential to a WT-like photoreaction, with a high accumulation of the late red-shifted intermediate P_3^{530}.

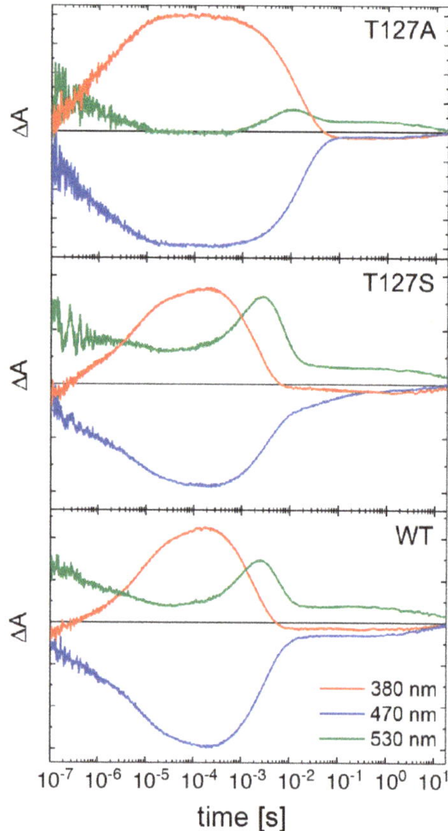

Figure 7. Kinetics in the UV/Vis range recorded after pulsed laser excitation (450 nm). Traces at 380 nm (red lines), 470 nm (blue), and 530 nm (green) are shown for each variant. While the T127A variant (top panel) showed a prolonged lifetime of the P_2^{380} state and reduced P_3^{530} formation, the T127S variant (middle panel) had very similar kinetics to the WT (bottom panel).

3.4. FTIR Difference Spectroscopy on the T127 Variants

FTIR spectroscopy was applied to gather information on the structural changes of the T127 variants (Figure 8). For comparison, difference spectra of *Cr*ChR2-WT were chosen at time points that represented mainly the intermediates P_2^{380} (300 μs) and P_3^{530} (6.7 ms), and were compared to the difference spectra of the T127A and T127S variants recorded at 8.4 ms after pulsed excitation. At this time, the P_2^{380} intermediate was predominant in the variant with minor contributions from P_3^{530} and P_4^{480} intermediates.

The infrared difference spectra basically confirmed the observations made in the visible spectral range (Figure 6). The low accumulation of the P_3^{530} state in the T127A variant was observed by comparing the difference spectrum at 8.4 ms with the one at 6.7 ms of the WT. The band at 1737 cm^{-1} assigned to the deprotonation of the proton donor to the Schiff base, D156 [16], was reduced in intensity, which indicated low accumulation of the deprotonated D156 species. This observation was expected as soon as the lifetime of the P_2^{380} state (with deprotonated Schiff base) was extended, as here in the T127 variant. The negative band at 1556 cm^{-1} in the difference spectra of WT was due to the ethylenic stretching vibration of the retinal in ground-state ChR2 [32]. This band was hardly seen in T127A due to spectral overlap by a positive contribution that is absent in the WT. Instead, a broad negative band at 1533 cm^{-1} was registered, which was a marker for the contribution of the P_4^{480} photoreaction [35].

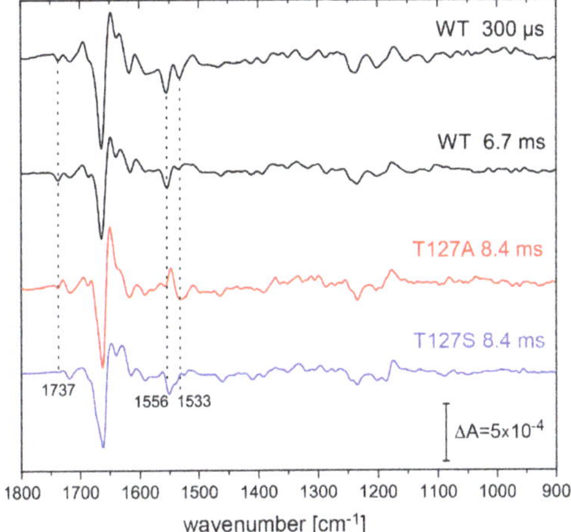

Figure 8. Time-resolved FTIR difference spectra of ChR2-WT (black spectra, taken from Reference [13]), T127A (red spectrum), and T127S variants (blue spectrum). The ChR2-WT spectrum at 300 μs and 6.7 ms was dominated by the P_2^{380} and P_3^{530} states, respectively. The spectra of the T127 variants were recorded at 8.4 ms, at which the P_2^{380} with smaller contributions of P_3^{530} was predominant.

4. Discussion

T127 of *Cr*ChR2 is localized in close vicinity to the retinal Schiff base and is hydrogen-bonded to E123 which, in turn, is the hydrogen bond acceptor of the Schiff base. The position of the threonine is also strategic for Schiff base reprotonation, as its hydroxylic group is supposed to be part of the pathway between the proton donor D156 and the Schiff base. There is no direct hydrogen-bonding network in the ground state that connects the $RSBH^+$ and D156, but it is only with the rise of the photocycle intermediate P_3^{530} that this connection is formed [13].

T127A substitution removed the hydroxylic group of the threonine, which affected the photocycle kinetics. The P_2^{380} intermediate, during the lifetime of which the Schiff base is deprotonated, had a ~10 times faster risetime, meaning that the deprotonation of the Schiff base was facilitated in this variant. The acceleration of the rise of the P_2^{380} intermediate may have been due to the stronger hydrogen-bonding interaction between the Schiff base and E123 that we reported in the present work.

The lifetime of the P_2^{380} state was longer in the T127A variant and, as a consequence, the accumulation of the subsequent intermediate P_3^{530} was very low. A longer lifetime of the state with a deprotonated Schiff base is a sign that the reprotonation pathway was blocked in T127A. The T127S variant conserved the hydroxyl group of the side chain, and we showed here that the photocycle kinetics were almost indistinguishable from the WT. We concluded, therefore, that the hydrogen-bonded network between D156 and the Schiff base, which is necessary for the reprotonation of the latter, involves the hydroxylic group of T127. The formation of this chain of hydrogen bonds is necessary for proton translocation and marks the transition between the P_2^{380} and P_3^{530} states.

T127 is important for the opening of the ion channel after light activation, since we have shown here that the T127A, as well as the more conservative T127S variants, showed less than 10% of the WT current in electrophysiology experiments. We note, though, that the reduced conductance may also have been related to lower expression yields of the variants in the different hosts (*Xenopus oocytes* and HEK cells). It has previously been shown that the channel closes with the decay of the P_3^{530} intermediate [36,37], but the channel opening is an optically silent transition in the visible range [38].

We confirmed from the present work that channel opening in *Cr*ChR2 is not related to changes in the photocycle kinetics as recorded by time-resolved UV/Vis spectroscopy. The T127S variant, in fact, showed WT-like photocyle kinetics, whereas a long-lived P_2^{380} intermediate occurred in the T127A variant with both variants exhibiting low channel conductance.

A threonine residue at this position was conserved for most of the cation- and anion-conducting channelrhodopsins. It is also present in BR [12,39], where exchanges to valine or alanine resulted in blue shifts in the absorption maxima by 146 nm and 28 nm, respectively, with minor effects on pumping activity [3]. By contrast, the exchange of threonine to serine and alanine in *Cr*ChR2 had only negligible effects on the electronic properties of the retinal chromophore, as registered by its visible absorption. This result is particularly interesting in light of the exceptionally high vibrational frequency recorded for the C=N–H vibration of the retinal Schiff base in T127A, as the hydrogen-bonding of the retinal Schiff base has been proposed to be a molecular determinant of the opsin shift [40].

QM/MM calculations showed nearly identical excitation energies for WT, T127S, and T127A variants. The same trend was also observed in the experimental absorption maxima of the retinal chromophore of *Cr*ChR2. However, a large shift of 19 cm^{-1} was calculated for the C=N–H vibration in the alanine variant, but no shift for the serine variant, again confirming the resonance Raman spectroscopic results. Since our QM/MM models were able to reproduce the relative trends in the UV/Vis and vibrational spectra of the variants, we considered these reliable for our analysis of the molecular changes. In the T127A variant, the hydrogen bond between the threonine and E123 was missing, affecting the hydrogen-bonding network around the Schiff base. To compensate for the missing hydrogen bond, the interaction between E123 and the Schiff base (N($RSBH^+$)$\cdots$ O(E123)) increased, resulting in a longer N–H bond which manifested in the upshift of the C=N–H vibrational frequency. The elongation of the Schiff base N–H bond was 0.016 Å, and possibly accelerates the deprotonation of the Schiff base in this variant. Hence, the stronger interaction between the Schiff base and E123 provided an explanation for the faster rise time of the P_2^{380} intermediate in the T127A variant, in a similar way to the D85E variant of bacteriorhodopsin [41].

Detailed analysis of the Schiff base vibration (coupled mode of the C=N stretching and the N–H bending vibration) by resonance Raman spectroscopy revealed the presence of two overlapping bands. The frequency shifts of these vibrational bands in the different variants were compared to the WT and their presence can be rationalized on the basis of the molecular model proposed in Reference [31]. In this model, the E123 side chain had one rotamer pointing towards the Schiff base and another rotamer pointing to a different hydrogen-bonding network that gave rise to two different vibrational bands of the C=N–H mode. This interpretation was supported by the band shape of the Schiff base vibration in the E123D variant, which was missing the high-frequency component. We can therefore lend support to the model of two rotamers of E123 in *Cr*ChR2, where a direct hydrogen bond with the Schiff base was formed in only one of the two configurations. The presence of two rotamer configurations may be essential for a voltage-sensing mechanism involving E123 [42]. While this seems to be a plausible scenario, further experimental evidence needs to be collected to support this model.

Author Contributions: Conceptualization, R.S. and J.H.; Funding acquisition, R.S. and J.H.; Investigation, D.E., N.K., M.S., C.B., R.K.K., K.H., D.H. and I.S.; Writing—original draft, R.S. and N.K.; Writing—review & editing, J.H. and M.S.

Funding: The research was funded by the German Research Foundation via the SFB-1078 projects B3 (to J.H.) and B4 (to R.S.). I.S. thanks the SFB 1078 for support within the Mercator program. R.K.K. acknowledges support from the Lady Davis Trust for Shunbrun postdoctoral fellowship. I.S. gratefully acknowledges funding by the European Research Council (ERC) under the European Union's Horizon 2020 research and innovation program (Grant No. 678169 "PhotoMutant").

Acknowledgments: The publication of this article was funded by Freie Universität Berlin.

Conflicts of Interest: The authors declare no conflict of interest.

References

1. Nagel, G.; Ollig, D.; Fuhrmann, M.; Kateriya, S.; Musti, A.M.; Bamberg, E.; Hegemann, P. Channelrhodopsin-1: A light-gated proton channel in green algae. *Science* **2002**, *296*, 2395–2398. [CrossRef] [PubMed]
2. Nagel, G.; Szellas, T.; Huhn, W.; Kateriya, S.; Adeishvili, N.; Berthold, P.; Ollig, D.; Hegemann, P.; Bamberg, E. Channelrhodopsin-2, a directly light-gated cation-selective membrane channel. *Proc. Natl. Acad. Sci. USA* **2003**, *100*, 13940–13945. [CrossRef] [PubMed]
3. Marti, T.; Otto, H.; Mogi, T.; Rosselet, S.J.; Heyn, M.P.; Khorana, H.G. Bacteriorhodopsin mutants containing single substitutions of serine or threonine residues are all active in proton translocation. *J. Biol. Chem.* **1991**, *266*, 6919–6927.
4. Rothschild, K.J.; He, Y.W.; Sonar, S.; Marti, T.; Khorana, H.G. Vibrational spectroscopy of bacteriorhodopsin mutants. Evidence that thr-46 and thr-89 form part of a transient network of hydrogen bonds. *J. Biol. Chem.* **1992**, *267*, 1615–1622. [PubMed]
5. Russell, T.S.; Coleman, M.; Rath, P.; Nilsson, A.; Rothschild, K.J. Threonine-89 participates in the active site of bacteriorhodopsin: Evidence for a role in color regulation and schiff base proton transfer. *Biochemistry* **1997**, *36*, 7490–7497. [CrossRef]
6. Pebay-Peyroula, E.; Rummel, G.; Rosenbusch, J.P.; Landau, E.M. X-ray structure of bacteriorhodopsin at 2.5 angstroms from microcrystals grown in lipidic cubic phases. *Science* **1997**, *277*, 1676–1681. [CrossRef]
7. Luecke, H.; Richter, H.T.; Lanyi, J.K. Proton transfer pathways in bacteriorhodopsin at 2.3 angstrom resolution. *Science* **1998**, *280*, 1934–1937. [CrossRef]
8. Kandori, H.; Kinoshita, N.; Yamazaki, Y.; Maeda, A.; Shichida, Y.; Needleman, R.; Lanyi, J.K.; Bizounok, M.; Herzfeld, J.; Raap, J.; et al. Structural change of threonine 89 upon photoisomerization in bacteriorhodopsin as revealed by polarized ftir spectroscopy. *Biochemistry* **1999**, *38*, 9676–9683. [CrossRef]
9. Kandori, H.; Yamazaki, Y.; Shichida, Y.; Raap, J.; Lugtenburg, J.; Belenky, M.; Herzfeld, J. Tight asp-85-thr-89 association during the pump switch of bacteriorhodopsin. *Proc. Natl. Acad. Sci. USA* **2001**, *98*, 1571–1576. [CrossRef]
10. Volkov, O.; Kovalev, K.; Polovinkin, V.; Borshchevskiy, V.; Bamann, C.; Astashkin, R.; Marin, E.; Popov, A.; Balandin, T.; Willbold, D.; et al. Structural insights into ion conduction by channelrhodopsin 2. *Science* **2017**, *358*, 8862. [CrossRef]
11. Hou, S.Y.; Govorunova, E.G.; Ntefidou, M.; Lane, C.E.; Spudich, E.N.; Sineshchekov, O.A.; Spudich, J.L. Diversity of chlamydomonas channelrhodopsins. *Photochem. Photobiol.* **2012**, *88*, 119–128. [CrossRef] [PubMed]
12. Govorunova, E.G.; Sineshchekov, O.A.; Rodarte, E.M.; Janz, R.; Morelle, O.; Melkonian, M.; Wong, G.K.-S.; Spudich, J.L. The expanding family of natural anion channelrhodopsins reveals large variations in kinetics, conductance, and spectral sensitivity. *Sci. Rep.* **2017**, *7*, 43358. [CrossRef] [PubMed]
13. Lorenz-Fonfria, V.A.; Resler, T.; Krause, N.; Nack, M.; Gossing, M.; Fischer von Mollard, G.; Bamann, C.; Bamberg, E.; Schlesinger, R.; Heberle, J. Transient protonation changes in channelrhodopsin-2 and their relevance to channel gating. *Proc. Natl. Acad. Sci. USA* **2013**, *110*, 1273–1281. [CrossRef] [PubMed]
14. Nack, M.; Radu, I.; Gossing, M.; Bamann, C.; Bamberg, E.; von Mollard, G.F.; Heberle, J. The dc gate in channelrhodopsin-2: Crucial hydrogen bonding interaction between c128 and d156. *Photochem. Photobiol. Sci.* **2010**, *9*, 194–198. [CrossRef] [PubMed]
15. Watanabe, H.C.; Welke, K.; Sindhikara, D.J.; Hegemann, P.; Elstner, M. Towards an understanding of channelrhodopsin function: Simulations lead to novel insights of the channel mechanism. *J. Mol. Biol.* **2013**, *425*, 1795–1814. [CrossRef] [PubMed]
16. Krause, N.; Engelhard, C.; Heberle, J.; Schlesinger, R.; Bittl, R. Structural differences between the closed and open states of channelrhodopsin-2 as observed by epr spectroscopy. *FEBS Lett.* **2013**, *587*, 3309–3313. [CrossRef]
17. Resler, T.; Schultz, B.J.; Lorenz-Fonfria, V.A.; Schlesinger, R.; Heberle, J. Kinetic and vibrational isotope effects of proton transfer reactions in channelrhodopsin-2. *Biophys. J.* **2015**, *109*, 287–297. [CrossRef]
18. Noguchi, T.; Sugiura, M. Flash-induced ftir difference spectra of the water oxidizing complex in moderately hydrated photosystem ii core films: Effect of hydration extent on s-state transitions. *Biochemistry* **2002**, *41*, 2322–2330. [CrossRef]

19. Schnedermann, C.; Muders, V.; Ehrenberg, D.; Schlesinger, R.; Kukura, P.; Heberle, J. Vibronic dynamics of the ultrafast all-trans to 13-cis photoisomerization of retinal in channelrhodopsin-1. *J. Am. Chem. Soc.* **2016**, *138*, 4757–4762. [CrossRef]
20. DeLano, W.L. *The Pymol Molecular Graphics System*; DeLano Scientific LLC: San Carlos, CA, USA, 2002.
21. Becke, A. Density-functional thermochemistry: The role of extract exchange. *J. Chem. Phys.* **1993**, *98*, 5648. [CrossRef]
22. Grimme, S.; Ehrlich, S.; Goerigk, L. Effect of the damping function in dispersion corrected density functional theory. *J. Comput. Chem.* **2011**, *32*, 1456–1465. [CrossRef] [PubMed]
23. Maier, J.A.; Martinez, C.; Kasavajhala, K.; Wickstrom, L.; Hauser, K.E.; Simmerling, C. Ff14sb: Improving the accuracy of protein side chain and backbone parameters from ff99sb. *J. Chem. Theory Comput.* **2015**, *11*, 3696–3713. [CrossRef] [PubMed]
24. Johnson, R. Nist Standard Reference Database Number 101. Available online: http://cccbdb.nist.gov (accessed on 1 October 2015).
25. Bannwarth, C.; Grimme, S. A simplified time-dependent density functional theory approach for electronic ultraviolet and circular dichroism spectra of very large molecules. *Comput. Theory Chem.* **2014**, *1040*, 45–53. [CrossRef]
26. Metz, S.; Kästner, J.; Sokol, A.A.; Keal, T.W.; Sherwood, P. C hem s hell—A modular software package for qm/mm simulations. *Wires Comput. Mol. Sci.* **2014**, *4*, 101–110. [CrossRef]
27. Radu, I.; Bamann, C.; Nack, M.; Nagel, G.; Bamberg, E.; Heberle, J. Conformational changes of channelrhodopsin-2. *J. Am. Chem. Soc.* **2009**, *131*, 7313–7319. [CrossRef] [PubMed]
28. Aton, B.; Doukas, A.G.; Callender, R.H.; Becher, B.; Ebrey, T.G. Resonance raman studies of the purple membrane. *Biochemistry* **1977**, *16*, 2995–2999. [CrossRef]
29. Nack, M.; Radu, I.; Schultz, B.J.; Resler, T.; Schlesinger, R.; Bondar, A.N.; del Val, C.; Abbruzzetti, S.; Viappiani, C.; Bamann, C.; et al. Kinetics of proton release and uptake by channelrhodopsin-2. *FEBS Lett.* **2012**, *586*, 1344–1348. [CrossRef]
30. Narva, D.; Callender, R.H. On the state of chromophore protonation in rhodopsin: Implication for primary photochemistry in visual pigments. *Photochem. Photobiol.* **1980**, *32*, 273–276. [CrossRef]
31. Guo, Y.; Beyle, F.E.; Bold, B.M.; Watanabe, H.C.; Koslowski, A.; Thiel, W.; Hegemann, P.; Marazzi, M.; Elstner, M. Active site structure and absorption spectrum of channelrhodopsin-2 wild-type and c128t mutant. *Chem. Sci.* **2016**, *7*, 3879–3891. [CrossRef]
32. Nack, M.; Radu, I.; Bamann, C.; Bamberg, E.; Heberle, J. The retinal structure of channelrhodopsin-2 assessed by resonance raman spectroscopy. *FEBS Lett.* **2009**, *583*, 3676–3680. [CrossRef]
33. Lórenz-Fonfría, V.A.; Schultz, B.-J.; Resler, T.; Schlesinger, R.; Bamann, C.; Bamberg, E.; Heberle, J. Pre-gating conformational changes in the cheta variant of channelrhodopsin-2 monitored by nanosecond ir spectroscopy. *J. Am. Chem. Soc.* **2015**, *137*, 1850–1861. [CrossRef] [PubMed]
34. Lorenz-Fonfria, V.A.; Bamann, C.; Resler, T.; Schlesinger, R.; Bamberg, E.; Heberle, J. Temporal evolution of helix hydration in a light-gated ion channel correlates with ion conductance. *Proc. Natl. Acad. Sci. USA* **2015**, *112*, 5796–5804. [CrossRef] [PubMed]
35. Saita, M.; Pranga-Sellnau, F.; Resler, T.; Schlesinger, R.; Heberle, J.; Lorenz-Fonfria, V.A. Photoexcitation of the p4(480) state induces a secondary photocycle that potentially desensitizes channelrhodopsin-2. *J. Am. Chem. Soc.* **2018**, *140*, 9899–9903. [CrossRef] [PubMed]
36. Bamann, C.; Kirsch, T.; Nagel, G.; Bamberg, E. Spectral characteristics of the photocycle of channelrhodopsin-2 and its implication for channel function. *J. Mol. Biol.* **2008**, *375*, 686–694. [CrossRef] [PubMed]
37. Ernst, O.P.; Murcia, P.A.S.; Daldrop, P.; Tsunoda, S.P.; Kateriya, S.; Hegemann, P. Photoactivation of channelrhodopsin. *J. Biol. Chem.* **2008**, *283*, 1637–1643. [CrossRef] [PubMed]
38. Lórenz-Fonfría, V.A.; Heberle, J. Channelrhodopsin unchained: Structure and mechanism of a light-gated cation channel. *Biochim. Biophys. Acta Bioenerg.* **2014**, *1837*, 626–642. [CrossRef] [PubMed]
39. Del Val, C.; Royuela-Flor, J.; Milenkovic, S.; Bondar, A.N. Channelrhodopsins: A bioinformatics perspective. *Biochim. Biophys. Acta* **2014**, *1837*, 643–655. [CrossRef] [PubMed]
40. Fodor, S.; Gebhard, R.; Lugtenburg, J.; Bogomolni, R.; Mathies, R. Structure of the retinal chromophore in sensory rhodopsin i from resonance raman spectroscopy. *J. Biol. Chem.* **1989**, *264*, 18280–18283.

41. Heberle, J.; Oesterhelt, D.; Dencher, N.A. Decoupling of photo- and proton cycle in the asp85–> glu mutant of bacteriorhodopsin. *EMBO J.* **1993**, *12*, 3721–3727. [CrossRef]
42. Berndt, A.; Schoenenberger, P.; Mattis, J.; Tye, K.M.; Deisseroth, K.; Hegemann, P.; Oertner, T.G. High-efficiency channelrhodopsins for fast neuronal stimulation at low light levels. *Proc. Natl. Acad. Sci. USA* **2011**, *108*, 7595–7600. [CrossRef] [PubMed]

Article

Light Stimulation Parameters Determine Neuron Dynamic Characteristics

Alexander Erofeev [1,*], Evgenii Gerasimov [1], Anastasia Lavrova [1,2,3], Anastasia Bolshakova [1], Eugene Postnikov [4], Ilya Bezprozvanny [1,5] and Olga L. Vlasova [1,*]

1 Graduate School of Biomedical Systems and Technologies, Laboratory of Molecular Neurodegeneration, Peter the Great St.Petersburg Polytechnic University, Khlopina St. 11, 194021 St. Petersburg, Russia
2 Medical Faculty, Saint-Petersburg State University, 7/9 Universitetskaya Emb., 199034 St. Petersburg, Russia
3 Saint-Petersburg Research Institute of Phthisiopulmonology, Polytechnicheskaya St. 32, 194064 St. Petersburg, Russia
4 Department of Theoretical Physics, Kursk State University, Radishcheva St. 33, 305000 Kursk, Russia
5 Department of Physiology, UT Southwestern Medical Center at Dallas, Dallas, TX 75390, USA
* Correspondence: alexandr.erofeew@gmail.com (A.E.); olvlasova@yandex.ru (O.L.V.)

Received: 8 August 2019; Accepted: 30 August 2019; Published: 5 September 2019

Featured Application: This report highlights the importance of light stimulation parameters (frequency, duration, intensity) on the activity of neurons expressing channelrhodopsin-2. These results will allow neuroscientists to stably activate neurons during a repeated light pulse train in optogenetic experiments with ChR2.

Abstract: Optogenetics is a recently developed technique that is widely used to study neuronal function. In optogenetic experiments, neurons encode opsins (channelrhodopsins, halorhodopsins or their derivatives) by means of viruses, plasmids or genetic modification (transgenic lines). Channelrhodopsin are light activated ion channels. Their expression in neurons allows light-dependent control of neuronal activity. The duration and frequency of light stimulation in optogenetic experiments is critical for stable, robust and reproducible experiments. In this study, we performed systematic analyses of these parameters using primary cultures of hippocampal neurons transfected with channelrhodopsin-2 (ChR2). The main goal of this work was to identify the optimal parameters of light stimulation that would result in stable neuronal activity during a repeated light pulse train. We demonstrated that the dependency of the photocurrent on the light pulse duration is described by a right-skewed bell-shaped curve, while the dependence on the stimulus intensity is close to linear. We established that a duration between 10–30 ms of stimulation was the minimal time necessary to achieve a full response. Obtained results will be useful in planning and interpretation of optogenetic experiments.

Keywords: membrane current; hippocampal neurons; optogenetics; light stimulation; channelrhodopsin-2

1. Introduction

Large scale neural activity patterns formed due to an interplay between dynamics of neurons and their network interconnections are principal fingerprints of brain functionality. At the same time, age and genetic disruption of neuronal connections can lead to neurodegenerative diseases such as Alzheimer's, Parkinson's, Huntington's and others. The number of incidence cases is increasing due to population ageing associated with new advances in medicine and technology. Therefore, the key problem is the difficult search for mechanisms of neurodegenerative disorders and new approaches to their treatment.

The most widespread conventional method for studying mechanisms of neuronal activity uses electrical stimulation of electrodes placed in the extracellular space. Electrophysiology allows for easy control of temporal resolution; nevertheless, in most cases it activates a lot of neurons simultaneously but not individual neurons. This is a significant disadvantage since specific types of neurons have an intrinsic activity pattern [1,2]. There is also an intracellular stimulation method which provides the necessary spatial and temporal resolutions, but its application is restricted to neuronal culture or brain slices [3].

One of the methods that allows for the study of the activity of certain neurons in vivo is the optical stimulation of neurons (optogenetics). In comparison with the methods mentioned above, it offers several advantages, such as a high spatio-temporal resolution with a parallel stimulation of certain brain areas [4,5]. The main principle of this method is based on the expression of light-sensitive ion channels called opsins in neuronal membranes. One of them, channelrhodopsin-2 (ChR2) isolated from the green alga *Chlamydomonas reinhardtii* [6,7], and its modifications is used practically for neuron excitation. ChR2 is activated with blue light (470 nm), which in turn induces a photoreceptor current due to non-selective cation flux into the cytoplasm of cells [6–8].

Since the method was developed, optogenetics has been widely used in neuroscience [4,8–11]. Recently, optogenetics has also been applied to investigate neurodegenerative diseases (an area of scientific interest of the authors). For example, the authors [12] used step-function opsin (SFO) for the long-term excitation of the hippocampal perforant pathway in amyloid precursor protein (APP) transgenic mice. As a result, a prolonged light excitation led to an increase in the level of amyloid deposits of peptide of 42 amino acid residues (Aβ42) that allows for the determination of a certain functional pathology in specific neuronal circuits in Alzheimer's disease. The review [13] contains a number of studies in which optogenetics is successfully used as a tool for activating or inhibiting specific regions or certain neurons involved in the development of neurodegenerative and neurological diseases such as Parkinson's disease [14], Huntington's disease [15], and epilepsy [16,17].

Despite the fact that optogenetics is one of the most developing areas of neurobiology and a widely used technique, there are some open questions related to the parameters of light stimulation for neurons expressing opsins. In this paper, we would like to emphasize the importance of light parameters during repeated light stimulation. Thus, our main goal is to determine the relationship between the light stimulation parameters (frequency, duration, intensity) and neuron activity during repeated light stimulation and define the optimal parameters for the stable activity of neurons.

2. Materials and Methods

2.1. Animals

The breeding colony of wild type mice of the same strain (C57BL/6J background, #000664) obtained from the Jackson Laboratory was established and maintained in a vivarium with 4–5 mice per cage and a 12 h light/dark cycle in the animal facility. All procedures were approved by principles of the European convention (Strasburg, 1986) and the Declaration of International Medical Association regarding the humane treatment of animals (Helsinki, 1996).

2.2. Hippocampal Primary Culture

The hippocampal cultures of mice were established from postnatal day 0–1 pups and maintained in culture as described earlier [18]. Briefly, after dissection and dissociation, neurons were plated on coverslips (pre-coated with poly-D-lysine, 0.1 mg/mL, #27964, Sigma, St. Louis, MI, USA) and cultured in neurobasal A medium with an addition of 1% fetal bovine serum (FBS) and 2% optimized neuronal cell culture serum-free supplement B27. On the third day of in vitro culture (DIV3), cytosine arabinoside (Ara-C) (40 M, Sigma, #C1768) was added to prevent the growth of glial cells. At DIV7 and DIV14, 50% of the medium was exchanged with fresh neurobasal A medium containing 2% B27 without FBS. At DIV7, neurons were transfected using the calcium phosphate method.

2.3. The Calcium Phosphate Method of Transfection

At DIV7, neurons were transfected using the calcium phosphate method with a mammalian transfection kit (#631312, Clontech Laboratories Inc., Mountain View, CA, USA) according to the manufacturer's protocol with recommendations from [19]. Due to the fact that the excitation length (470 nm) of green fluorescent protein (GFP) coincides with the excitation length of chanalrhodopsin-2, we used the marker plasmid pCSCMV:tdTomato (#30530, Addgene, Watertown, MA, USA), which encodes a red fluorescent protein with an excitation wavelength of 530 nm. For multiple plasmid delivery, we used the calcium phosphate co-transfection method. Plasmids were mixed at a ratio of 3:1, with 3 parts containing the plasmid of interest.

2.4. Whole Cell Patch Recordings in Hippocampal Cultures

The recording of neuronal light-induced activity was performed with whole-cell voltage- and current-clamp techniques after 14 days of cultivation; at this time, neurons mature and form a stable network [20,21]. Whole-cell recordings were conducted in artificial cerebrospinal fluid (ACSF) external solution, which contains 124 mM NaCl, 26 mM $NaHCO_3$, 10 mM glucose, 5 mM KCl, 2.5 mM $CaCl_2$, 1.3 mM $MgCl_2$ and 1 mM NaH_2PO_4 (pH 7.4, 290–310 milliosmole (mOsm)) [22]. An Olympus IX73 inverted microscope was used with the 40× objective and appropriate filters for fluorescence imaging. This microscope was equipped with a 4-channel light-emitting diode (LED) driver Thorlabs DC4104 to localize and target fluorescent ChR2-expressing neurons by expression of the tdTomato fluorescent protein.

Patch pipettes, pulled from borosilicate glass (1.5 mm outside diameter (OD), 0.86 mm inside diameter (ID), 3-5 MOm) were filled with a solution containing 140 mM K-Gluconate, 2 mM $MgCl_2$, 2 mM NaCl, 2 mM ATP-Na_2, 0.3 mM GTP-Mg, 10 mM 4-(2-hydroxyethyl)-1-piperazineethanesulfonic acid (HEPES) (pH 7.35, 290–300 mOsm). Conventional whole-cell patch-clamp recordings were obtained via a MultiClamp 700 B double patch amplifier and a Digidata 1440A (Molecular Devices, Sunnyvale, CA, USA) coupled to the acquisition software pClamp 10.7. The holding potential in voltage-clamp mode was −70 mV, uncorrected for any liquid junction potential (in the range of −10 mV to −8 mV) between the internal and external solutions. After DIV14, light-evoked neuronal activity has been detected using the whole-cell patch clamp technique as shown in Figure 1. We have used 10 primary cultures for the measurement of membrane currents and potentials.

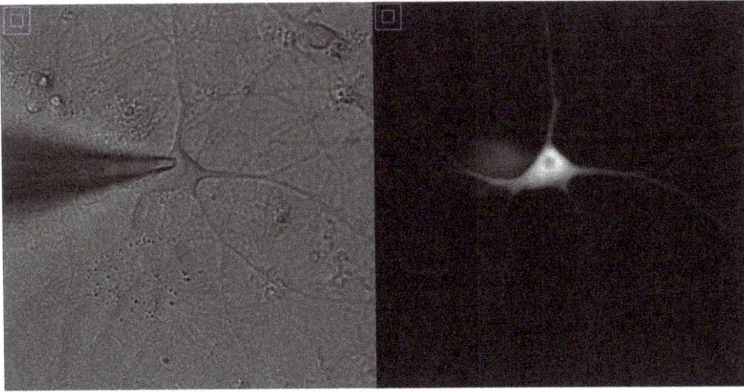

Figure 1. Primary hippocampal neuron transfected with FCK-ChR2-GFP and pCSCMV:tdTomato during the recording of photocurrents.

2.5. Optogenetic Method

During the recording of neuronal activity, a pulse train of blue light (470 nm) was performed with different parameters (frequency, duration, intensity). In our experiments, we used a frequency range of 1–30 Hz in increments of 5 Hz and a pulse duration range of 1–5 ms in increments of 1 ms, 10–50 ms in increments of 10 ms, and 100–500 ms in increments of 100 ms. Ten light stimuli were used per frequency or duration (Figure 2).

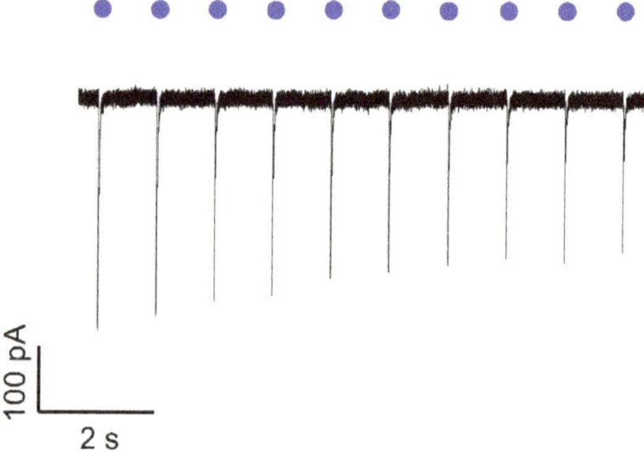

Figure 2. Photocurrents recorded using the whole-cell voltage-clamp technique at maximal intensity and light duration. t = 100 ms.

The maximum intensity of the blue LED (LED4D067, 470 nm, Thorlabs Inc., Newton, NJ, USA) was 35 mW mm^{-2} with a maximum photo flux of 250 mW. The values of light intensities were normalized, taking the maximum value in each sequence as unity, i.e., all indicated values are shown as relative ones.

The LED source has a modulation voltage in the range of 0–10 V and a current in the range of 0–1000 mA. We measured the intensity at the maximum current, i.e., 1000 mA, which corresponds to a modulation voltage of 10 V. To change the intensity, we connected a pulse generator to the LED source and changed the modulation voltage from 0 V to 10 V; while we considered that at a voltage of 10 volts, we get the maximum intensity. The values of light intensities were normalized, taking the maximum value in each sequence as unity.

2.6. Statistical Analysis

The mathematical software OriginPro 9.0 and GraphPad Prism 7 were used to process the data and obtain standard errors.

3. Results

To define the relationship between the activity of the pyramidal neuron expressing ChR2 and the parameters of the light stimulus, we carried out experiments with a 10-light pulse train. We picked up the frequency at which the action potential (AP) was generated for each light stimulus (Figure 3). We considered the generation of action potentials because the membrane current changed at all frequencies used in this study.

These results demonstrated a reduced number of AP with increasing frequency under blue light pulsing. The frequency at which the AP was generated for each light stimulus was in the range of 1–5 Hz. Further measurements were carried out at a frequency of 1 Hz. These results are similar to the data published previously [23], but the range of optimal stimulation frequency was more narrow in our experiments. The reason for this discrepancy requires further study.

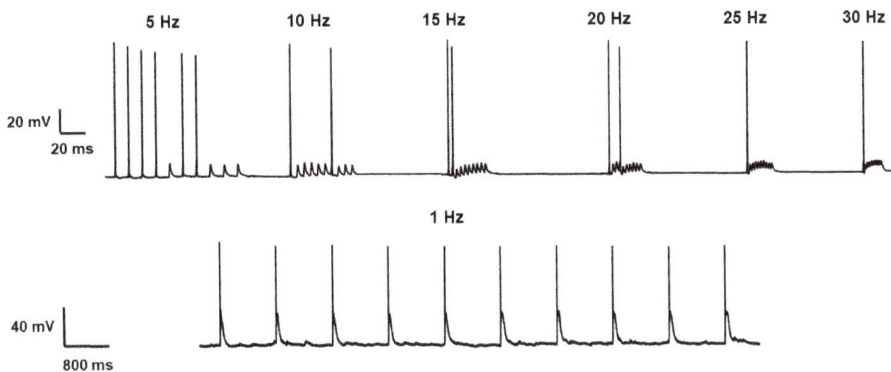

Figure 3. Action potential generation under blue light stimulation at different frequencies.

We have studied the effect of the light pulse duration in the range of 1–5 ms, 10–50 ms and 100–500 ms on the amplitude of photocurrents at various LED intensities; the value of maximum intensity has been normalized to unity. Each point in Figure 4 represents the mean value of the measured amplitudes of the response to a 10-pulse train with fixed duration and intensity (see Figure 2).

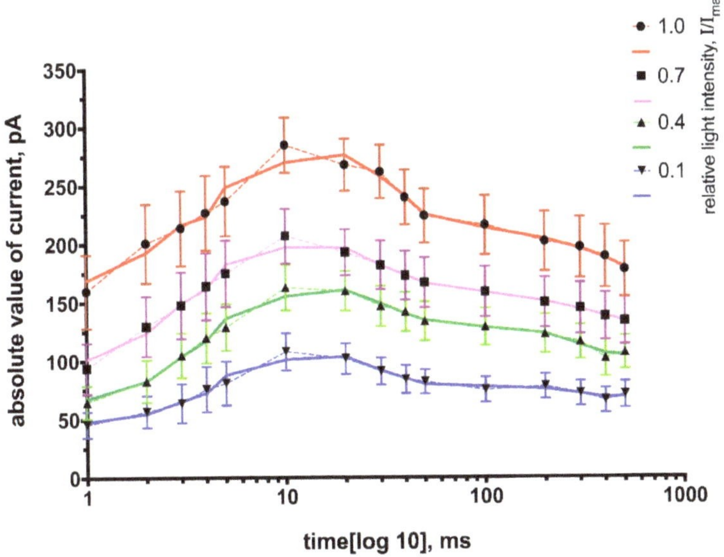

Figure 4. Dependence of the membrane current amplitude for pyramidal neurons on the duration of light stimulation at different light intensities (the value of intensity maximum is taken as 1), time–light pulse duration, $n = 10$. Plots of mean values with error bars indicate the standard error (SE). To obtain these mean values, fourth-order polynomial smoothing over neighboring points was performed in GraphPad Prism 7.

As shown in Figure 4, the dependency of the photocurrent amplitude on the light pulse is described by a right-skewed bell-shaped curve. The obtained dependency was identical for all light intensities used in the experiment. The obtained results agree with results published earlier [7,24].

It should be noted that in the range of 1–20 ms, the value of this amplitude increases, while at higher values of pulse duration (30–500 ms), the amplitude decreases. An explanation for this fact may be based on the kinetic model of the ChR2 photocycle, in which three channel states (open (O), desensitized (D) and closed (C)) were considered [7]. The three states model was the first model describing a photocycle of channelrhodopsin-2. However, after the spectral analysis of ChR2 [25], an existence of four kinetic intermediate states (P_1, P_2, P_3 and P_4) was indicated. Based on this information, four [24,26] and six [27] state models have been developed. To explain our findings, we utilized the four-state model of ChR2 photocycle, which consists of two open states ($O_1 \rightarrow O_2$) and two closed states ($C_1 \rightarrow C_2$). Thus, the states defined by the spectral analysis can be interpreted as follows: $O_1 \rightarrow O_2$ corresponds to the states P_2 and P_3 (open states with the time constants of 1 ms and 10 ms, respectively) and state C_2 (the time constant ~5s) corresponds to the state P_4 (desensitized). The state of C is the ground state of the channel and corresponds to P_0. We suppose that the decrease of photocurrents after 30 ms is due to the transition in the P_4 state, i.e., inactivation of ChR-2.

We have also studied the time interval tau (τ), which is needed to reach the maximum photocurrent (Figure 5) for different pulse durations (t) atmaximum intensities (I_{max}).

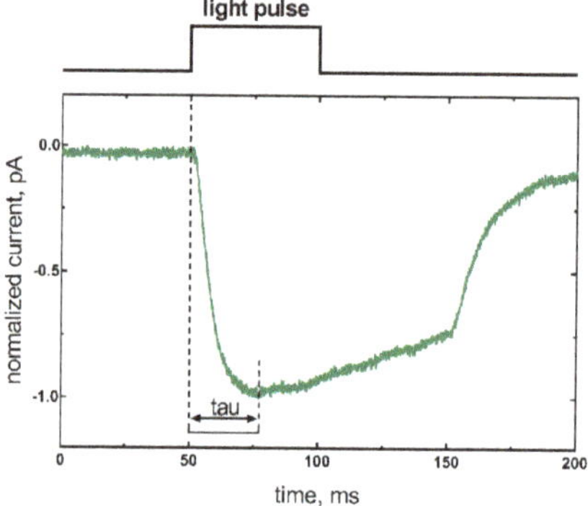

Figure 5. Light-induced currents with a pulse duration of 100 ms.

We analyzed the relationship between τ and light pulse duration (t) (Figure 6). According to dependence in the range of 10–30 ms, τ corresponds to the light pulse duration.

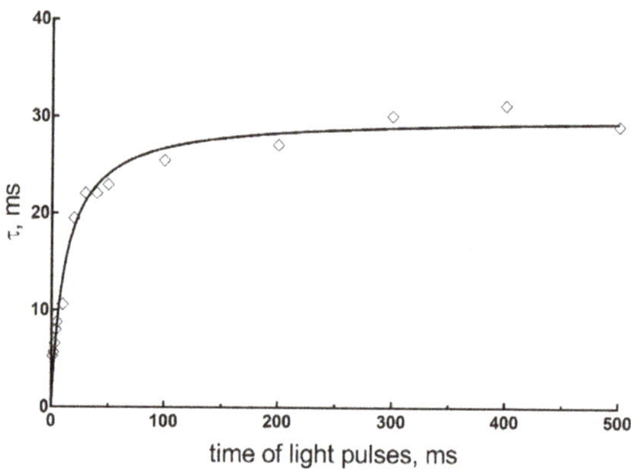

Figure 6. Dependence of τ on the duration of light pulses.

After t = 50 ms, the time τ did not change. This suggests that light stimulation with a duration longer than 50 ms is impractical, because the response of neurons expressing ChR2 is not stable.

Figure 7 shows an example of ChR2 photocurrents of the primary hippocampal neuron for the first light stimuli at various pulse durations (1–5 ms, 10–50 ms, 100–500 ms).

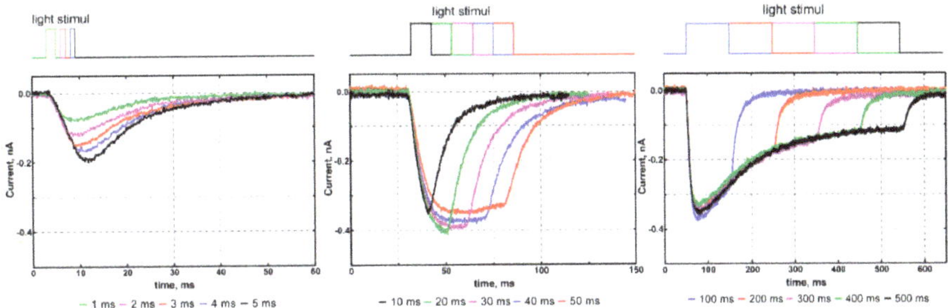

Figure 7. Light-induced currents recorded from one hippocampal neuron at different light pulse durations in the voltage-clamp mode of the whole-cell configuration: 1–5 ms (**left**), 10–50 ms (**center**), 100–500 ms (**right**). Light intensity is at the maximum value.

As shown in Figure 7 (left panel), the maximum photocurrent is already achieved after the light stimulus itself in the case of short pulse durations. At higher pulse durations (Figure 7, middle panel), τ either coincides with the pulse duration, or achieves maximum value before the end of the light stimulus itself. At large values of pulse duration (Figure 7, right panel), the time interval τ is always shorter than the duration of the light stimulus.

In the next series of experiments, we studied the effect of light intensity on photocurrent amplitude at different light pulse durations.

The obtained curves (Figure 8) demonstrate a mostly linear dependency that is closed to the generation of sodium currents [28].

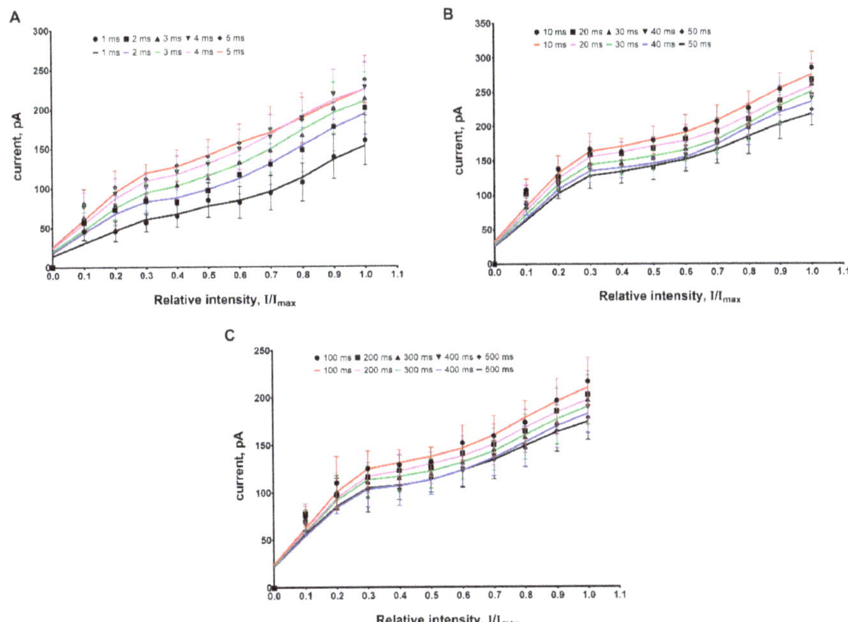

Figure 8. Dependence of the current amplitude (the average value for 10 light stimuli) for hippocampal neurons on the intensity of exposure at different light pulse durations: (**A**) 1–5 ms, (**B**) 10–50 ms, (**C**) 100–500 ms, $n = 10$. Plots of mean values with error bars indicate the standard error (SE). To obtain these mean values, fourth-order polynomial smoothing over neighboring points was performed in GraphPad Prism 7.

The obtained dependency of photocurrent amplitude on light intensity corresponds with previously published data [23].

4. Discussion

In this study, we analyzed a range of ChR2 stimulation conditions in experiments with primary hippocampal cultures.

We have determined an optimal frequency of light stimuli for generating APs in the current-clamp recording configuration. The optimal frequency range was between 1–5 Hz. Stimulation with frequencies less that 1 Hz was insufficient to generate APs, and stimulation with frequencies over 5 Hz results in the loss of fidelity of responses.

We have observed that the amplitude of ChR2 currents depends non-linearly (right-skewed bell-shaped curve) on the pulse duration. ChR2 photocurrent amplitudes were stable in the range of 10–30 ms, however, at low (1–5 ms) and large (100–500 ms) pulse duration values, the amplitudes changed in an almost stochastic manner (Figure 2).

Amplitude differences at low (1–5 ms) values of pulse duration are presumably explained due the fact that ChR2 only reaches the O_1 state and cannot change to O_2 with maximum bandwidth. At the same time, the observed amplitude differences at large pulse durations are explained due to the desensitization and degradation of ChR2 and the P_4 state [7]. Thus, we can conclude that the optimal interval of light stimuli is in the range of 10–30 ms.

The relationship between the photocurrent amplitude and the intensity of the light stimulus can be explained by the fact that increasing intensity leads to an increase of O_2 channel state duration. At intensities above 35 mW mm^{-2}, the amplitude of the photocurrent of ChR2 may decrease; this assumption needs further research.

Thus, we found that the following parameters of light stimulation: frequency (F) = 1–5 Hz, t = 10–30 ms, $I \leq I_{max}$ are optimal for multiple light stimulation, i.e., the activity of neurons expressing ChR2 will be stable throughout the stimulation period. This is also shown in Figure 9.

Figure 9 demonstrates the cumulative effect of duration and intensity on the activity of neurons expressing ChR2. Thus, the photocurrent amplitude is almost constant for the pulse duration, which is equal to 10 ms in the full range of intensities, while there is a wide variation of the amplitude values at other durations (1 ms and 300 ms) and intensities.

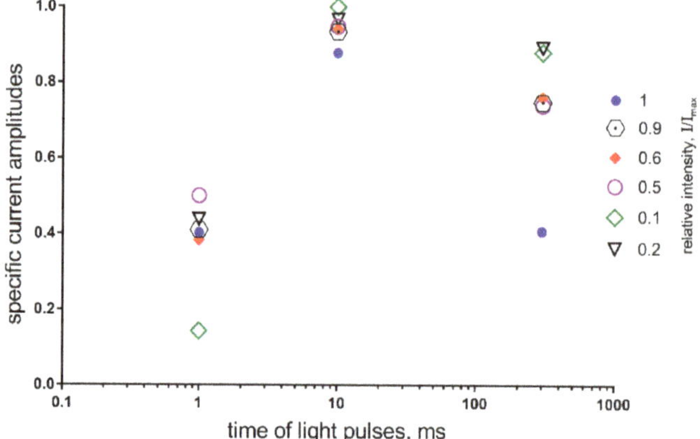

Figure 9. Graph representing the relative difference between the first two current spike amplitudes as a function of the light impulse duration and their relative intensity. The maximal difference is normalized to unity; the relative light intensities are indicated by markers. This graph was designed as follows: the chosen pulse duration was assigned to a value of the difference between amplitudes of the first and second pulses.

5. Conclusions

In this study, we determined the relationship between the light stimulation parameters (frequency, duration, intensity) and neuron activity during repeated light stimulation and defined the optimal parameters for the stable activity of neurons. It was determined that the optimal frequency of light stimuli for generating APs is in the range of 1–5 Hz. We demonstrated that the dependency of the current amplitude on light pulse duration is described by a right-skewed bell-shaped curve, while the dependence on stimulus intensity is close to linear. We discovered that the light pulses between 10–30 ms in the full range of intensity are optimal for activation of ChR2 in cultured hippocampal neurons. We established that a 10 ms duration of stimulation was the minimal time necessary to achieve full response. The obtained results will be useful in the planning and interpretation of optogenetic experiments.

Author Contributions: Conceptualization, O.L.V. and A.E.; methodology, O.L.V.; validation A.L. and O.L.V.; formal analysis, A.E., E.G. and A.L.; investigation, A.E. and E.G.; resources, A.B. and I.B.; writing—original draft preparation, A.E., A.L. and O.L.V.; writing—review and editing, E.P. and A.B.; visualization, A.E.; supervision, O.L.V.; project administration, I.B.; funding acquisition, I.B. and E.P.

Funding: This research was funded by the state grant 17.991.2017/4.6 (IB) and by the Russian Science Foundation Grant No. 19-15-00184 (IB). The financial support was divided in the following way: experiments depicted in Figures 1–6 were supported by the state grant 17.991.2017/4.6, experiments depicted in Figures 6 and 7 were supported by the Russian Science Foundation Grant No. 19-15-00184, and experiments in Figures 8 and 9 were supported by the Russian Science Foundation, project 19-15-00201 (AL and EP).

Conflicts of Interest: The authors declare no conflict of interest.

References

1. Hutcheon, B.; Yarom, Y. Resonance, oscillation and the intrinsic frequency preferences of neurons. *Trends Neurosci.* **2000**, *23*, 216–222. [CrossRef]
2. Somogyi, P.; Klausberger, T. Defined types of cortical interneurone structure space and spike timing in the hippocampus. *J. Physiol.* **2005**, *562*, 9–26. [CrossRef] [PubMed]
3. Sakmann, B.; Neher, E. Patch Clamp Techniques for Studying Ionic Channels in Excitable Membranes. *Annu. Rev. Physiol.* **1984**, *46*, 455–472. [CrossRef] [PubMed]
4. Callaway, E.M.; Yuste, R. Stimulating neurons with light. *Curr. Opin. Neurobiol.* **2002**, *12*, 587–592. [CrossRef]
5. Gradinaru, V.; Thompson, K.R.; Zhang, F.; Mogri, M.; Kay, K.; Schneider, M.B.; Deisseroth, K. Targeting and Readout Strategies for Fast Optical Neural Control In Vitro and In Vivo. *J. Neurosci.* **2007**, *27*, 14231–14238. [CrossRef] [PubMed]
6. Nagel, G.; Ollig, D.; Fuhrmann, M.; Kateriya, S.; Musti, A.M.; Bamberg, E.; Hegemann, P. Channelrhodopsin-1: A light-gated proton channel in green algae. *Science* **2002**, *296*, 2395–2398. [CrossRef] [PubMed]
7. Nagel, G.; Szellas, T.; Huhn, W.; Kateriya, S.; Adeishvili, N.; Berthold, P.; Ollig, D.; Hegemann, P.; Bamberg, E. Channelrhodopsin-2, a directly light-gated cation-selective membrane channel. *Proc. Natl. Acad. Sci. USA* **2003**, *100*, 13940–13945. [CrossRef] [PubMed]
8. Glock, C.; Nagpal, J.; Gottschalk, A. Microbial rhodopsin optogenetic tools: Application for analyses of synaptic transmission and of neuronal network activity in behavior. In *Methods in Molecular Biology*; Humana Press: Totowa, NJ, USA, 2015; Volume 1327, pp. 87–103. ISBN 1588295974.
9. Zhang, H.; Cohen, A.E. Optogenetic Approaches to Drug Discovery in Neuroscience and Beyond. *Trends Biotechnol.* **2017**, *35*, 625–639. [CrossRef] [PubMed]
10. Mahmoudi, P.; Veladi, H.; Pakdel, F.G. Optogenetics, Tools and Applications in Neurobiology. *J. Med. Signals Sens.* **2017**, *7*, 71–79. [PubMed]
11. Rost, B.R.; Schneider-Warme, F.; Schmitz, D.; Hegemann, P. Optogenetic Tools for Subcellular Applications in Neuroscience. *Neuron* **2017**, *96*, 572–603. [CrossRef]
12. Yamamoto, K.; Tanei, Z.; Hashimoto, T.; Wakabayashi, T.; Okuno, H.; Naka, Y.; Yizhar, O.; Fenno, L.E.; Fukayama, M.; Bito, H.; et al. Chronic Optogenetic Activation Augments Aβ Pathology in a Mouse Model of Alzheimer Disease. *Cell Rep.* **2015**, *11*, 859–865. [CrossRef] [PubMed]

13. Vann, K.T.; Xiong, Z.G. Optogenetics for neurodegenerative diseases. *Int. J. Physiol. Pathophysiol. Pharmacol.* **2016**, *8*, 1–8. [PubMed]
14. Gradinaru, V.; Mogri, M.; Thompson, K.R.; Henderson, J.M.; Deisseroth, K. Optical deconstruction of parkinsonian neural circuitry. *Science* **2009**, *324*, 354–359. [CrossRef] [PubMed]
15. Cepeda, C.; Galvan, L.; Holley, S.M.; Rao, S.P.; André, V.M.; Botelho, E.P.; Chen, J.Y.; Watson, J.B.; Deisseroth, K.; Levine, M.S. Multiple sources of striatal inhibition are differentially affected in Huntington's disease mouse models. *J. Neurosci.* **2013**, *33*, 7393–7406. [CrossRef] [PubMed]
16. Soper, C.; Wicker, E.; Kulick, C.V.; N'Gouemo, P.; Forcelli, P.A. Optogenetic activation of superior colliculus neurons suppresses seizures originating in diverse brain networks. *Neurobiol. Dis.* **2016**, *87*, 102–115. [CrossRef]
17. Tonnesen, J.; Sorensen, A.T.; Deisseroth, K.; Lundberg, C.; Kokaia, M. Optogenetic control of epileptiform activity. *Proc. Natl. Acad. Sci. USA* **2009**, *106*, 12162–12167. [CrossRef]
18. Popugaeva, E.; Pchitskaya, E.; Speshilova, A.; Alexandrov, S.; Zhang, H.; Vlasova, O.; Bezprozvanny, I. STIM2 protects hippocampal mushroom spines from amyloid synaptotoxicity. *Mol. Neurodegener.* **2015**, *10*, 37. [CrossRef]
19. Jiang, M.; Chen, G. High Ca2+-phosphate transfection efficiency in low-density neuronal cultures. *Nat. Protoc.* **2006**, *1*, 695–700. [CrossRef]
20. Biffi, E.; Regalia, G.; Menegon, A.; Ferrigno, G.; Pedrocchi, A. The influence of neuronal density and maturation on network activity of hippocampal cell cultures: A methodological study. *PLoS ONE* **2013**, *8*, e83899. [CrossRef]
21. Grabrucker, A.; Vaida, B.; Bockmann, J.; Boeckers, T.M. Synaptogenesis of hippocampal neurons in primary cell culture. *Cell Tissue Res.* **2009**, *338*, 333–341. [CrossRef]
22. Zhang, H.; Liu, J.; Sun, S.; Pchitskaya, E.; Popugaeva, E.; Bezprozvanny, I. Calcium Signaling, Excitability, and Synaptic Plasticity Defects in a Mouse Model of Alzheimer's Disease. *J. Alzheimer's Dis.* **2015**, *45*, 561–580. [CrossRef] [PubMed]
23. Ishizuka, T.; Kakuda, M.; Araki, R.; Yawo, H. Kinetic evaluation of photosensitivity in genetically engineered neurons expressing green algae light-gated channels. *Neurosci. Res.* **2006**, *54*, 85–94. [CrossRef] [PubMed]
24. Hegemann, P.; Ehlenbeck, S.; Gradmann, D. Multiple photocycles of channelrhodopsin. *Biophys. J.* **2005**, *89*, 3911–3918. [CrossRef] [PubMed]
25. Bamann, C.; Kirsch, T.; Nagel, G.; Bamberg, E. Spectral Characteristics of the Photocycle of Channelrhodopsin-2 and its Implication for Channel Function. *J. Mol. Biol.* **2008**, *375*, 686–694. [CrossRef] [PubMed]
26. Nikolic, K.; Grossman, N.; Grubb, M.S.; Burrone, J.; Toumazou, C.; Degenaar, P. Photocycles of channelrhodopsin-2. *Photochem. Photobiol.* **2009**, *85*, 400–411. [CrossRef] [PubMed]
27. Grossman, N.; Simiaki, V.; Martinet, C.; Toumazou, C.; Schultz, S.R.; Nikolic, K. The spatial pattern of light determines the kinetics and modulates backpropagation of optogenetic action potentials. *J. Comput. Neurosci.* **2013**, *34*, 477–488. [CrossRef] [PubMed]
28. Halterman, M.W. Neuroscience. In *Neurology*, 3rd ed.; Sinauer Associates, Inc.: Sunderland, MA, USA, 2004; Volume 64, p. 43.

Review

Two-Photon Excitation of Azobenzene Photoswitches for Synthetic Optogenetics

Shai Kellner and Shai Berlin *

Department of Neuroscience, Ruth and Bruce Rappaport Faculty of Medicine, Technion-Israel Institute of Technology, Haifa 3525433, Israel; shaikellner@gmail.com
* Correspondence: shai.berlin@technion.ac.il

Received: 1 December 2019; Accepted: 20 January 2020; Published: 23 January 2020

Abstract: Synthetic optogenetics is an emerging optical technique that enables users to photocontrol molecules, proteins, and cells in vitro and in vivo. This is achieved by use of synthetic chromophores—denoted photoswitches—that undergo light-dependent changes (e.g., isomerization), which are meticulously designed to interact with unique cellular targets, notably proteins. Following light illumination, the changes adopted by photoswitches are harnessed to affect the function of nearby proteins. In most instances, photoswitches absorb visible light, wavelengths of poor tissue penetration, and excessive scatter. These shortcomings impede their use in vivo. To overcome these challenges, photoswitches of red-shifted absorbance have been developed. Notably, this shift in absorbance also increases their compatibility with two-photon excitation (2PE) methods. Here, we provide an overview of recent efforts devoted towards optimizing azobenzene-based photoswitches for 2PE and their current applications.

Keywords: two-photon; azobenzene; optogenetics; photoswitch

---

1. Introduction

Photoreceptors are protein light-detectors that allow the cell/organism to respond to its environment [1]. Most, if not all, photoreceptors are a two-component system: a genetically-encoded protein coupled to a chromophore. Following photon-absorption, chromophores undergo unique alterations (for instance in geometry or charge [2,3]), resulting in the modulation of the protein partner to which each chromophore is bound to. In most instances, this process is completely reversible, whereby removal of light (i.e., return to darkness) facilitates the relaxation of the chromophore to its initial, thermally stable state. This also restores the function of the protein back to as it were prior illumination. These features, explicitly non-invasive reversible control over proteins, have made it possible to develop fantastic genetically-encoded optical tools—denoted 'optogenetic' tools—for the manipulation of defined subsets of cells [4]. Many tools today are based on naturally-occurring photoreceptors, for instance opsins [5–8]. However, the need to photocontrol other signaling mechanisms, proteins or molecules, have pushed forward the development of complementary strategies, explicitly Synthetic optogenetics [9,10].

Synthetic optogenetics relies on the use of non-natural, synthetic chromophores—photoswitches—to relay photon absorption to modulation of structure and function of their target molecule (Figure 1a). Photoswitches do not require a unique chromophore binding pocket or protein domains as in the case of chromophores (e.g., [11]). Instead, photoswitches can be directly conjugated to various locations on their target molecule with use of different genetic or chemical handles (see below) [9,12]. However, unlike chromophores that can be synthesized by various cells (even if the cells are not intrinsically light-sensitive), photoswitches need to be supplemented to the preparation. This is both a blessing and a curse. Blessing—since the preparation remains naïve until the photoswitch is added. Curse—it adds another layer of complexity to the experiment, especially in vivo, such as injection prior experimentation.

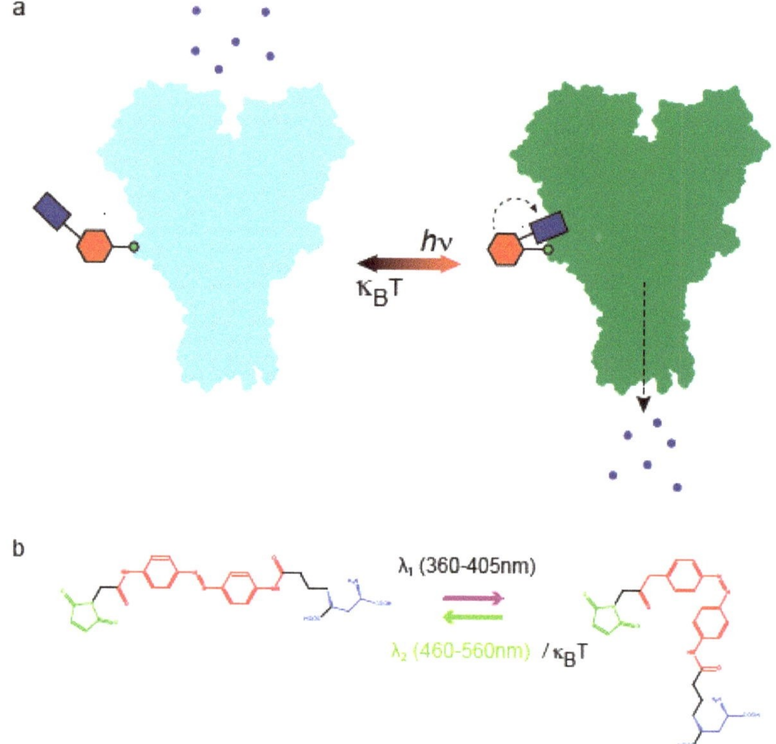

Figure 1. Synthetic optogenetics. (**a**) Cartoon depiction of a photoswitchable glutamate receptor (space filling based on PDB: 3KG2). For simplicity, only two of four subunits are presented. In addition, only one subunit is shown to tether the photoswitch (color-coded shapes). The photoswitch consists of a ligand (purple rectangle), an azobenzene-core (red hexagon) and a maleimide for attachment (green circle). In the dark, the photoswitch is in *-trans* and channel is closed (leftward cartoon, light blue). Photoisomerization of the photoswitch to *-cis* by red-shifted light ($h\nu$) allows insertion of the glutamate headgroup (purple rectangle) into its binding pocket inducing channel opening (rightward cartoon, green receptor; dashed arrow depicts conductance of ions). (**b**) Chemical structure of an exemplary photoswitch; MAG (color coded as in (a); maleimide—green; azobenzene—red; glutamate—purple). Visible light photoisomerization (common to most photoswitches): 360–405 nm light photoisomerizes MAG from *-trans* to *-cis*; illumination at ~460–560 nm (or thermal relaxation; $\kappa_B{}^T$) returns MAG to *-trans*.

Photoswitches can be custom-tailored to photomodulate a large variety of effectors. For instance, some have been fashioned to mechanically restructure membrane lipids or DNA molecules [13–15], to act as light-activated forceps [16] or to include pharmacological agents (e.g., pore-blockers) to block, agonize, or antagonize receptors, ion channels, or enzymes (Figure 1b) (also see reviews [9,17,18]). It is noteworthy to mention that although this method dates back to the late 1960s (Figure 2) [19], it has particularly gained momentum in early 2000 owing to critical advances in chemistry, biology, and imaging methods. Here, we focus on the most recent efforts devoted to push it to the next level, namely towards multiphoton activation of photoswitches for in vivo applications in opaque preparations.

Owing to the versatility of the technique, we foresee a bright future for 2PE-compatible photoswitches in biology and medicine.

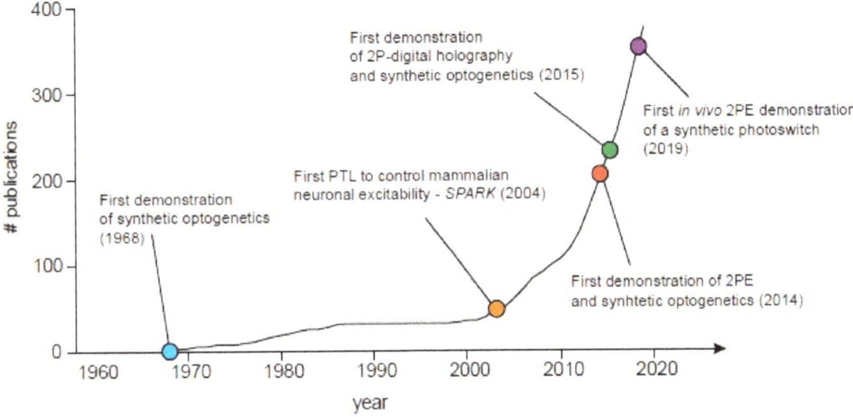

Figure 2. Graphical illustration of 'synthetic optogenetics' emerging in the scientific literature. Cumulative plot of number of publications across years was performed by searching for specific terms (chemical-optogenetics, photopharmacology, optopharmacology, synthetic optogenetics, optogenetic pharmacology, photoswitch, azobenzene and photo), as well as by searching for publications from recognized researchers in the field (e.g., Erlanger BF, Poolman B, Feringa B, Lester HA, Trauner D, Kramer RH, Isacoff EY, Woolley GA, Gorostiza P, Paoletti P, Ellis-Davies GCR, etc.). Search was continuously performed in various search engines, notably PubMed Central, Google Scholar, and bioRxiv. Last query was performed on November 2019.

2. Photoswitches

At the heart of photoswitches lies a light-sensitive core [20]. Of the different kinds, the most prevalent moiety used in biological applications is the azobenzene (Figure 1b) [21]. Its widespread use arises from favorable features, such as high quantum yield, minimal photobleaching, and fast responsivity at biologically-relevant conditions, namely pH, temperature and solubility in water-based solvents [21–23]. However, its ability to undergo *trans*-to-*cis* photoisomerization is undoubtfully the main reason; giving rise to robust changes in geometry and end-to-end distance of the molecule (Figure 1b).

Azobenzene photoswitches commonly include additional chemical groups placed on either side of the light-sensitive core. These can be quite diverse, such as drugs, chemical- or biological tethers, dyes, as well as lipidic structures [9]. The nature of these groups will dictate the mode of action of the photoswitch and whether it is to remain diffusive or immobilized to its target. Diffusive photoswitches (photochromic ligands; PCL) are, in principle, reversible caged-compounds [24]. In contrast, immobilized photoswitches (photoswitchable tethered ligand; PTL [25]) include a unique chemical tether that can conjugate specific residues of the protein (e.g., maleimide to bind cysteines [26]; benzylguanine to a SNAP domain [27]). These restrict the binding (thereby the effect) of the photoswitch to a defined molecule. Importantly, specific amino acids (a.a.) or domains can be genetically-encoded, thereby giving users access to defined populations of cells. Today, most targets for PTLs require genetic modifications, with cysteines as the preferred choice. However, recent advances have made it possible to conjugate photoswitches to endogenous, non-modified receptors by use of strong electrophilic moiety in the PTL to couple with reactive amines and hydroxyl groups, naturally present in several a.a. side chains [28]. Another scheme to bypass protein modification is by attaching the photoswitch next to its intended target. This may include expression of membrane anchors (e.g., a transmembrane domain with the tether facing the extracellular) to conjugate the photoswitch which can then modulate adjacent endogenous proteins [29].

3. Photomodulating Cellular Activity

One prototypical example of a synthetic optogenetic tool is SPARK (Synthetic Photoisomerizable Azobenzene-Regulated K^+-channel) [30]. SPARK was designed to control neuronal excitability. This channel was designed to tether MAL-AZO-QA. MAL-AZO-QA includes an azobenzene (AZO) flanked by a cysteine-reactive maleimide (MAL) and a potassium channel blocker (quaternary ammonium; QA). For the maleimide to specifically tether SPARK, the channel had a cysteine residue inserted in one of its external loops (as water-accessible cysteines are not common in proteins [31]). Prior illumination, the photoswitch is found in its elongated *-trans* form; spanning 17 Å. The authors designed the length of the photoswitch to equal the distance between the cysteine residue and the pore so that the QA drug would easily reach the pore and block the channel. The functional outcome of blocking this potassium channel is robust neuronal excitability and action potential firing. This effect was completely reversed by near-UV illumination (~400 nm), pushing the photoswitch back to *-cis*, a much shorter form (end-to-end distance ~10 Å), thereby physically removing the blocker away from the pore leading to immediate silencing of the neuron [30]. This could be repeated many times, by quickly toggling the photoswitch back to *-trans* by ~500 nm light (rather than slow thermal relaxation in the dark). Similar schemes have been adopted to photocontrol a variety of channels and receptors with the, almost exclusive, use of visible light [9,12].

4. Shifting from Visible to Near-Infrared

Despite progresses made since the very first demonstrations of Synthetic optogenetics by the Erlanger group in the late 1960s [19,32], current illumination schemes remain as they were back then, namely rely on the use of visible light, typically near-UV (~400 nm) for *trans-to-cis* photoisomerization and blue-shifted wavelengths (~500) for *cis-to-trans* [9,12,15,17,18,24,33,34]. Aside the potential cytotoxicity of these wavelengths [35,36], visible light is not well-suited for deep tissue penetration due to strong absorption and scatter from endogenous components [37]. The use of longer, near infrared (NIR; >700 nm) wavelengths is advantageous as these are less absorbed by biological tissues and water [38], induce less photodamage [36], and are less prone to scatter.

Synthetic optogenetics is slow to adapt NIR illumination owing to the very poor absorbance of these low energy wavelengths by the azobenzene core, a common challenge with other optogenetic tools [39], not to mention the relatively little information extant for NIR absorption by photoswitches ([40,41] and see below). In fact, though the *-cis* azobenzene was first discovered almost a century ago (see [42]), the debate regarding the exact mechanism for photoisomerization, and how different factors affect it, remains lively to this day [43,44]. Following excitation, the azobenzene molecule proceeds from S0 to S1 and S2 states, with distinct absorption bands for each transition (n–π^* transition excite azobenzene compounds to S1 state and π–π^* transition leads to S2 state) [45,46]. However, many different factors can have dramatic influences on the these, for instance substitutions on the phenyl rings, solvent properties, temperature, to name the more common factors. However, it is much less known by which of the proposed mechanisms the isomerization of azobenzene proceeds, namely rotation, inversion, concerted inversion, and inversion assisted rotation (for more details see [44]). These uncertainties arise from different approaches used, or different experimental settings. For instance, whereas non-viscous polar solvents favor a rotation mechanism, viscous non-polar favor inversion [44]. It has also been shown that the use of different wavelengths can affect the isomerization mechanism. In fact, these differences are also seen with other compounds such as fluorescent proteins [47,48] and extant data of how wavelengths affect isomerization mostly pertains to one-photon excitation of azobenzene compounds [49]. This actuality makes it very difficult to infer the mechanisms for NIR excitation as the absorption properties between these different illumination schemes are substantially different [41]. For instance, use of 1-photon for isomerization shows that the transition from *trans*-to-*cis* path develops triexponentially with times of 0.3, 3, and 16 ps (where the first two reflect population relaxation and the third corresponds to the final relaxation to ground state [49]) and these are not affected by solvent viscosity rather by wavelengths at early (t~1 ps) and late (t~100 ps) times but show similar photoisomerization behavior on a 10 ps time scale [50].

Some of these hurdles can be overcome by using, at least, two NIR photons; each with half the energy of needed, to be absorbed in a single event (within ~10^{-18} s) in order to reach the molecule's excited state [51]. For instance, if a photoswitch/chromophore efficiently absorbs near-UV light at 400 nm, in principle it should be possible to excite it by two simultaneous photons at approximately 800 nm each [41]. This technique is denoted two-photon excitation (2PE) [52]. However, it is important to note that 2PE is a third-order nonlinear process that depends not only on the absorption cross-section (σ2) of the molecule, but also on the concentration of incident photons. Therefore, even small differences in laser conditions can lead to very different absorptions; making it difficult to compare compounds, or even the relative strength of transitions. Nevertheless, recent reports addressing these issues show that the fundamental excitation of the S0→S2 transition by 2P are similar and are due to resonance absorption [50].

2PE allows to excite molecules found deeper within the tissue (up to mm), at high three-dimensional, sub-micrometric spatial resolution [51,53]. This provides exquisite means to activate optogenetic tools at the level of single cells, or even subcellular regions, with minimal light escaping to undesired or nearby regions [54]. However, many optical tools display low 2PE absorption cross-section. This is worsened by low expression of the protein under control or its low conductance (in the case of an ion channel) [55,56]. Together, these limitations largely render 2PE unusable in these instances. Improving 2PE could be obtained by increasing the 2PE cross-section absorption of a photoswitch by chemical modification (s) (see below), by increasing the expression and of the protein or to simultaneously illuminate larger regions-of-interest (ROIs) such as entire somata. Expansion of the illumination area can be obtained by parallel excitation techniques that provide a high flux of photons to numerous ROIs simultaneously, in contrast to rastering methods (line scanning) that provide photons to single pixels sequentially [55,57]. A collection of these improvements have been employed for 2P-photolysis of caged-compounds as well as photoactivation of various opsins for quite some time now [56,58–62] and, only recently, in synthetic optogenetics (Figure 2).

5. Two-Photon Compatible MAG-Based Photoswitches

One of the most commonly employed tethered photoswitch is MAG [9] (Figures 1b and 2). MAG, similar to MAL-AZO-QA (see above), has an azobenzene-core (A), flanked by a cysteine-reactive maleimide (M), but instead of a channel blocker bears a glutamate molecule (G). This photoswitch is therefore intended for glutamate-binding proteins. The MAG photoswitch has been employed to activate glutamate receptors (e.g., LiGluR, LiGluN, mGluRs) in its *-cis* form, not to mention to antagonize a glycine-binding receptor (the GluN1a-subunit [26]). Akin to other azobenzene-based photoswitches, MAG undergoes very efficient *-trans* to *-cis* photoisomerization when irradiated with near-UV light (MAG_0; $\lambda_{trans\text{-}cis}$ 340–400 nm), and reverts from *-cis* back to *-trans* by thermal relaxation or, significantly more rapidly, by blue-green wavelengths (MAG_0; $\lambda_{cis\text{-}trans}$ 440–580 nm). Nevertheless, with regards to 2PE, MAG exhibits very poor absorption in the red-to-NIR range (~700–1400 nm) [41].

In the case of first generation MAGs (MAG0 [26,63,64]), endowed with a symmetrically-substituted azobenzene (Figure 2 and Table 1), the 2P-absorption cross section is low (σ2 = 10 GM at 820 nm [41]), albeit on the order of magnitude of a common fluorescent protein such as eGFP (σ2 = 30 GM at 927 nm [47]). This, along the lower expression and density of channels at membrane of neurons, results in very little or no capacity to photocontrol cellular activity when 2PE is employed [41]. This prompted the design of second-generation MAGs with increased 2P-absorption cross sections. More precisely, this required lowering the energetic barrier of *trans-to-cis* isomerization by specifically adjusting and shifting MAG's spectral properties towards 'red'-er wavelengths.

Table 1. Applications of 2PE for synthetic optogenetics.

Photoswitch	Type	1P and 2P Peak Absorption (nm)	2P-Absorption Cross-Section (GM)	Two Photon Imaging Method	Cell Type	Application	Ref
L-MAG_0	PTL	1P 376 nm 2P maximal responses at 820 nm	σ2 = 10 GM at 820 nm [a] and σ2 = 300 GM at 630 nm (-*trans*) [a]	2P-laser scanning at 820 nm	HEK293	2PE elicited optimal photocurrents of I_{max}~120 pA in GluK2-L439C-(LiGluR) expressing HEK293 cells. Responses were 10–20% of those obtained by 1P.	[65]
MAG2p	PTL	1P ~420 nm 2P maximal response at 900 nm	σ2 = 56 GM at 780 nm [b]	2P-laser scanning at 900 nm	HEK293 and dissociated hippocampal neurons	Cells expressing LiGluR, in voltage clamp 2PE elicited photocurrents recordings of ~50 and ~20 pA in HEK and neurons, respectively. These were 10–20% of those obtained by 1P illumination.	[65]
MAGA2p	PTL	1P ~420 nm 2P maximal response at 880 nm	Naphthalene (antenna), σ2 ≈ 200 GM at 780 nm [c]	2P-laser scanning at 900 nm	HEK293	Cells expressing LiGluR, in voltage clamp 2PE elicited photocurrents recordings of 6 pA.	[65]
MAGA ligand-2	PTL	1P ~380 nm	Pyrene (antenna), σ_2 = 55 GM	Not tested under 2PE	N.A.	N.A.	[66]
L-MAG_{460}	PTL	1P ~460 nm	σ2 = 80 GM at 850 nm (-*trans*)	2P digital holography (2P-DH) at 850 nm	HeLa cells and dissociated hippocampal neurons	Photocurrents elicited by 2P-DH were ~80 pA in size, ~39% of those obtained by 1P photocurrents from the same cell. HeLA cells expressing R-GECO showed reliable fluorescent responses when imaged at 561 nm and excited at 840 nm.	[41]
D-MAG_{460}	PTL	Similar properties as L-MAG_{460}		2P digital holography (2P-DH) At 850 nm	HEK293 cells	Cells expressing mGluR3-Q306C (LimGluR3) illuminated by 2P-DH exhibited mean photocurrents of ~100 pA, namely ~85% of those obtained by 1P.	[41]
$MAG^{slow}_{2p\_F}$	PTL	1P ~360 nm	σ2 = 69 GM [d]	2P-laser scanning at 780 nm	HEK293 cells, rat hippocampal organotypic slices and *C. elegans*' TRN neurons	Cells expressing LiGluR and R-GECO or RCaMP showed that 2PE elicited photoresponses (Δ*F*/*F*) similar to those obtained by 1P illumination (at 405 nm). Similar photoresponses from 1P and 2P were seen in vivo, in TRN neurons in *C. elegans* expressing GCaMP6s.	[67]
toCl-MAG1	PTL	1P ~470 nm	N.A.	Not tested under 2PE	N.A.	N.A.	[68]
ATG	PCL	1P ~330 nm	N.A.	2P-laser scanning at 740 nm	Hippocampal neurons	*cis*-ATG-mediated synaptic currents of ~50 pA in size following 2P-photolysis.	[69]
PAI	PCL	1P ~320 nm	N.A.	2P-laser scanning at 840 nm	HEK293 cells	Cells co-expressing mAChR (m2R) and Gαq(TOP) loaded with the calcium indicator Oregon-green (OGB-AM) displayed reliable fluorescent responses (ΔF/F) to *trans*-PAI by 2PE.	[70]
Alloswitch	PCL	1P and 2P absorbance N.A. 2P maximal response obtained by 760 nm	N.A.	2P-laser scanning	HEK cells and organotypic hippocampal slices.	HEK cells expressing mGluR5-eYFP loaded with Fura-2 AM or neurons in organotypic hippocampal slices expressing GCaMP6s and DsRed were photostimulated by 2PE. *trans*-alloswitch released local inhibition of mGlu5 and enabled DHPG (mGluR5-agonist)-dependent activation of the receptor. Responses were monitored using Ca^{2+}-indicators. 760 nm elicited highest increase in the frequency of calcium-oscillations.	[71]
Glu_brAzo2	PCL	1P -*cis* ~395 nm-*trans* ~480 nm	N.A.	Not tested under 2PE	N.A.	N.A.	[72]
4FAB-QA	PCL	1P ~415 nm	N.A.	2P-laser scanning at 780 and 1000 nm	CA1 neurons from hippocampal organotypic slices	2PE elicited ~32% inhibition of voltage gated Na^+ and K^+ channels than that obtained by 1P.	[73]

[a]; Data from Carroll et al. [41]. [b]; Data from Cabré et al. [67]. [c]; Data from Gascón-Moya et al. [66]. [d]; Data estimated based on density functional theory, N.A = not available.

One strategy employed involved breaking the symmetry around the azobenzene core (below in Figure 3 and Table 1) [74]. This was obtained by creating a push–pull system, where one benzene ring of the azo-core was decorated with an electron donating group, while the second azo-unit was supplemented with an electron withdrawing group. These led to an asymmetric azobenzene with a significantly red-shifted absorbance [21,68,74–77]. This strategy was employed in one of the very first 2P-compatible PTLs, where the MAG photoswitch was redesigned to include an asymmetric aminoazobenzene core (tertiary amine in the 4′-position) to act as a strong electron-donating group, denoted MAG2p (Figure 3) [65]. Indeed, this photoswitch exhibited a red-shifted 1P-absorption spectrum peaking at 420 nm; ~60 nm red-shifted compared to the parent MAG. Notably, this photoswitch could be activated by 2PE (Table 1). In a subsequent report, the authors estimated its 2P-absopriton cross section and suggest it to be ~five times higher than that of the parent photoswitch (MAG2p, $\sigma 2$ = 56 GM [72] at 850 nm, Table 1). The authors go on to show that, when coupled to a glutamate receptor (LiGluR [63]) optimal 2P-responses are obtained at 900 nm. Another unique feature of this photoswitch is that it is no longer bi-stable, rather spontaneously reverts back to *-trans* in the dark. Notably, decrease in the thermal stability of the *-cis* isomer by red-shifting of absorbance of azobenzenes is a well-described phenomenon [74]. The functional outcome of the thermal relaxation of the photoswitch is a rapid cessation of channel activation, seen as a decrease in current once illumination stops (τ_{off} = 150 ms [67]; ~10,000-fold times faster than thermal relaxation of MAG, τ_{off} = 25 min [75]).

A push–pull scheme was similarly employed, very shortly after, by Kienzler et al. [75]. Here, the authors have modified MAG to include an asymmetric azobenzene-core with a tertiary amine at the 4′-position as the electron-donor, but an acetamido-group at the 4′-position, as the weak electron withdrawing group (Figure 3). These modifications resulted in a larger shift in peak 1P-absorption (~100 nm) of the *-trans* isomer (λ_{peak} = 460 nm), therefore denoted MAG_{460} (Figure 3) [75]. Expectedly, this modification reduced the stability of the *-cis* isomer (MAG_{460}, τ_{off} = 0.71 s) [75]. Importantly, this shift led to an 8-fold increase in the photoswitch's 2P-absorption cross section (*trans*-MAG_{460}, $\sigma 2$ = 80 GM at 850 nm) [41]. We have later employed this photoswitch to successfully activate LiGluR [41]. Though the photocurrents obtained by 2PE were of smaller amplitude than when using 1P-illumination, they were sufficient to evoke action potential firing in cultured neurons by parallel excitation methods (digital-holography; Table 1) [41,78]. Lastly, we also employed a fluorescent reporter as readout for Ca^{2+}-activity (R-GECO) and find it to provide reliable responses ($\Delta F/F$), demonstrating that this method is also suitable for all-optical interrogation of cells [78].

Another interesting strategy that has been employed to sensitize the MAG photoswitch towards NIR wavelengths is by adding a light-harvesting molecule (i.e., antenna). Two such photoswitches were designed under the name of maleimide–azobenzene–glutamate antenna, or MAGA. The first—MAGA2p [65]—consisted of the same asymmetric core as us found in MAG2p along an added light-harvesting naphthalene-derivative (with a high 2P-absorption cross section; $\sigma 2 \approx 200$ GM at 780 nm [79]) (Figure 3). The antenna was incorporated to sensitize the *trans*-azobenzene by resonant electronic energy transfer (RET) (Figure 3). This photoswitch could therefore be photoactivated by two distinct mechanisms: direct excitation of the push–pull azo-core by 2PE and RET by the photosensitizing antenna. This photoswitch displayed low thermal stability, with rapid relaxation of the currents (τ_{off} = 265 ms [65]) that, when excited using a laser-scanning method, would not allow efficient activation of the channel or robust cellular responses. In a subsequent improved version, MAGA ligand-2 (Figure 3) [66], the authors maintained an antenna for harvesting light, albeit swapped it by a pyrene molecule (σ_2 = 55 GM), but reverted back to a symmetrical azo-core with the intention to counteract the pyrene's lower 2P-absorption cross section (~4-fold lower than naphthalene) by increasing the thermal stability of the photoswitch. Indeed, the resulting photoswitch exhibited a highly stable *-cis* isomer (MAGA ligand-2, τ_{off} = 2.0 h in DMSO). Thus, the increased thermal stability of the *-cis* state should make this photoswitch more suitable to accumulate activated photoswitches by scanning imaging methods. However, it is worthy to mention that the photosensitizing antenna employed drastically reduced the solubility of the photoswitch and, likely, its usability in biological experiments. Indeed, so far MAGA ligand-2 was not employed with cells (Table 1).

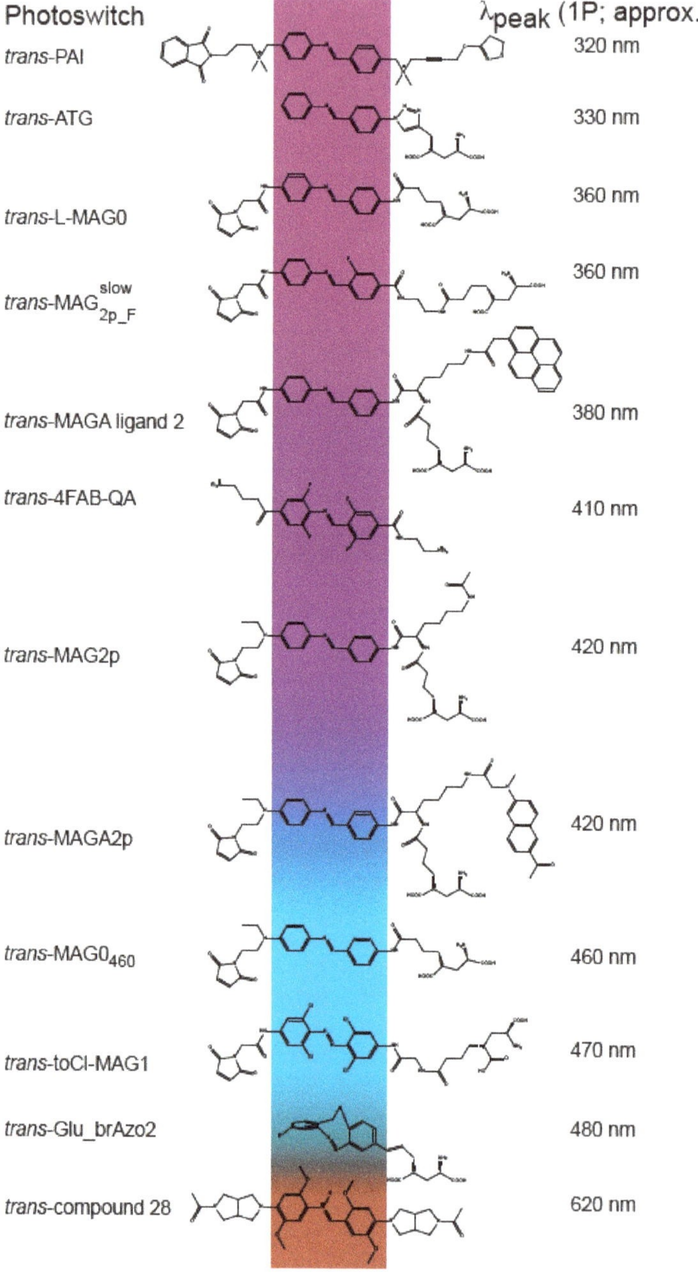

Figure 3. Chemical structure and absorption of exemplary 2P-compatible photoswitches. Left, names of photoswitches; right, one photon (1P) peak absorption (λ_{max}), also depicted by middle color gradient.

The most recent development in MAG photochemistry includes a compromise between the -*cis*' thermal stability and 2P-absorptivity by addition of a strong inductive electron withdrawing group in the *ortho* position, in conjunction with an asymmetric azo-core (Figure 3). This new design has been based on previous *ortho* substitutions made onto azobenzene molecules that can affect the steric or electronic barrier for isomerization, not to mention to robustly slow down its thermal relaxation rates [68,74]. Examples of this include the symmetrical azo-core photoswitch denoted toCl-MAG1 (Figure 3); highly decorated with tetra-*ortho*-chloro substitutions, exhibiting a very slow thermal relaxation (τ_{cis} = 3.5 h at 37 °C in PBS, pH = 7.0) [68], not to mention red-shifting its peak 1P absorption to 470 nm (the most rightward shift seen in PTLs) (Figure 3). When computationally modeled onto the azo-core of MAG0 and MAG2p, these substitutions yielded new photoswitches, designated $MAG^{slow}_{2p\_F}$ (Figure 3) [67]. Surprisingly, these substitutions did not lead to a major shift in the 1P absorption of the photoswitch (λ_{peak} = ~360 nm), however they significantly increased the thermal stability of the -*cis* variant when compared to previous MAG2p (MAG2p, τ_{cis} = 118 ms in 80% PBS:20% DMSO [65]; $MAG^{slow}_{2p\_F}$, τ_{cis} = 10 min in 99% PBS:1% DMSO [67]). Modeling also showed that the photoswitch should exhibit a slightly increased 2P-absoprtion cross section (MAG2p, $\sigma2$ = 56 GM; $MAG^{slow}_{2p\_F}$, $\sigma2$ = 69 GM), therefore potentially resolving the problem of insufficient accumulation of opened channels during illumination. Although the authors do not show comparison of 1P vs. 2P-mediated current size (which should differ), they do however extensively characterize the responses obtained by a fluorescent calcium activity reporter (RCaMP2) and show that equivalent responses are obtained when using optimal 1P-(405 nm) or 2P-excitation (780 nm) (Table 1). These results are in-line with our own, showing the compatibility of this photoswitch, and technique, for all-optical interrogation of cells.

6. Novel Photoswitches with 2PE-Potential

Other azobenzene-based photoswitches have also been rendered 2p-compatible by use of other strategies. Of these, the methoxy-substitutions at the *meta* positions and C2-bridged azobenzene are particularly interesting [74]. Methoxy *meta*-substitutions are estimated to strongly shift the absorbance of the azo-core to much longer wavelengths than most reported photoswitches (i.e., >100 nm), specifically into the far-red and infra-red regimes [80] and these should likely have a high 2P-absoprion cross section. However, these modifications will most likely lead to very rapid thermal relaxations. One such example is compound 28 with near-IR absorbance peaking at 620 nm (but also sufficiently activated by 730 nm), but with an ultra-rapid thermal back reaction (τ_{cis} = 10 µs) (Figure 3) [74].

The second approach, the C2-bridged azobenzene, also show a significant red-shifted absorption spectrum of the -*trans* isomer (λ_{max} > 540 nm) [72,74,81], implying an increased 2P-absorption cross section [41], but with a much more favorable (i.e., slower) thermal relaxation rate (on the order of minutes at room temperature in aqueous solution) than methoxy *meta*-substitutions. The drawback of the molecule is that the -*cis* state is the more thermodynamically stable form. This means that if the photoswitch is active in -*cis*, it would be active when applied onto the preparation. Thus, it is preferable should the photoswitch be active only in -*trans*. An additional drawback is that the stable -*cis* isomer poorly absorbs red light and isomerization to -*trans* requires UV to near UV illumination. Recently, two *trans*-active diffusible photoswitches were designed based on a C2-bridged azobenzene, denoted locked-azobenzene (LAB) [82] or Glu_brAzo [72]. Both photoswitches bear a glutamate moiety as ligand, thus intended for activation of glutamate receptors, and exhibit highly similar structural and spectral properties (Figure 3). Of note, despite their similarities, LAB is shown to selectively activate NMDARs, whereas Glu_brAzo efficiently activates both Kainate and AMPA receptors. Regardless the receptor type, Glu_brAzo shows more practical features for use in vivo, particularly increased solubility in aqueous media owing to the addition of a bulky ionic group. This is noteworthy as solubility issues are a very common limitation of most photoswitches, in particular azobenzene-based ones [9,21]. This additional group is also suggested to improve the performance of the photoswitch by increasing the steric congestion around the glutamate moiety with the intention to lower the interaction of the glutamate moiety with the receptor when the photoswitch is found in its stable -*cis* isomeric form. To design Glu_brAzo, the authors searched for *trans*-active photoswitches that target glutamate receptors. This was not very challenging as most diffusible GluR-photoswitches are, surprisingly, active in the extended thermally stable -*trans* form (e.g., [83,84]). They honed-in on a *trans*-active photoswitch denoted GluAzo [85]. For photoactivation, *cis*-Glu_brAzo (i.e., inactive) requires λ_{peak} = 395 nm to shift to -*trans*. As noted above, this is quite a limitation for two-photon activation. However, this photoswitch exhibits very desirable bi-stability, with very slow spontaneous back-isomerization from *trans*-to-*cis* ($t_{1/2}$~4 h, at RT in aqueous media). This step could be accelerated by green light illumination (λ_{peak} = 480 nm), suggesting that it may also be compatible with 2PE. However, the 2P-absorption cross section was not assessed for Glu_brAzo (or LAB) and is very hard to predict based solely on 1P absorbance (seen above and e.g., [47]). An additional limitation of Glu_brAzo is its overlapping absorption spectra of both -*cis* and -*trans* isomers, so much so that even the use of optimal 1P wavelengths result in a mixture of photostationary states, with merely ~60% of the active -*trans* isomer.

7. Future Prospects

There is growing interest in synthetic optogenetic tools in fields such as in neuroscience [8,12,23], heart physiology [86–88] and, intuitively, vision restoration [89–92]. Thus, we expect these to motivate further developments of the strategy in the upcoming years, in particular towards progression of the technique towards enhancing multiphoton absorption. Synthetic optogenetic tools can be combined with electrophysiological recordings and optical tools (e.g., Ca^{2+}-probes [78]), making it highly tailored for in vivo use in different animal models in an all-optical manner. We also suggest that several photoswitches might even make it to the clinic [93], in particular photoswitches that target native proteins (e.g., [28,29]) or those applicable in the blood [74]. In addition, synthetic optogenetic tools have shown promise for studying (and maybe treating) Parkinson's disease-related receptors [94], inhibiting cellular division in cancer [95], not to mention controlling cellular excitability, highly relevant for brain diseases such as epilepsy [96].

Here, we have briefly summarized several of the newest developments in the field of synthetic optogenetics, more precisely efforts devoted towards rendering photoswitches absorbent of longer, red-shifted wavelengths and their ability to undergo 2PE. These properties are highly desired for making the method more operational in vivo. Despite this motivation, it remains that most laboratories focus on designing unique photoswitches for 1P-applications. We believe that this stems from limited information on nonlinear multiphoton properties of photoswitches and the difficulties in determining, *a priori*, their multiphoton absorption properties [47,67], relaxation rates and whether optimization (derivatization) of 1P-functional photoswitches towards 2PE will not render them inoperative.

It appears that bistable photoswitches provide an optimal starting point for designing next-generation 2PE-compatible photoswitches. The slow relaxing photoswitches with light-harvesting antennae and symmetrical azo-cores display high 2P-absorption cross section that should allow for robust photoactivation of optical actuators in vivo. However, they will require further modifications to increase their solubility in aqueous solution prior use with cells. The bistable $MAG^{slow}_{2p\_F}$ exhibits slightly better features. Though of lower 2P cross-section (<70 GM) and almost no shift in 1P absorption maxima, its lower thermal relaxation allows for larger responses. The effect of this, and other, photoswitches could be further improved by additional means such as use of higher 2P-laser intensities (keeping in mind that higher intensities may be harmful to cells [52]), increase in the expression of the optical actuators [97,98] and, importantly, use of parallel excitation methods for simultaneous activation of larger regions of interest in all axes [55,57,78,99].

In summary, we see current hurdles as wonderful incentives and opportunities for designing better photoswitches and protein actuators. Indeed, many labs around the world are working intensely to address these issues. Consequently, we anticipate that the potential of synthetic optogenetics towards in vivo use and the clinic will be realized within the next few years.

Author Contributions: Conceptualization, S.B. and S.K.; methodology, S.B.; data curation, S.B. and S.K.; writing—review and editing, S.B. and S.K.; visualization, S.B.; supervision, S.B.; funding acquisition, S.B. The research submitted is in partial fulfilment for a Doctoral degree for S.K. All authors have read and agreed to the published version of the manuscript.

Funding: This research was funded by the Israel Science Foundation (ISF), grant number 1096/17 (S.B.).

Conflicts of Interest: The authors declare no conflict of interest.

References

1. Kottke, T.; Xie, A.; Larsen, D.S.; Hoff, W.D. Photoreceptors take charge: Emerging principles for light sensing. *Ann. Rev. Biophys.* **2018**, *47*, 291–313. [CrossRef] [PubMed]
2. Mathes, T.; Kennis, J.T.M. Editorial: Optogenetic tools in the molecular spotlight. *Front. Mol. Biosci.* **2016**, *3*. [CrossRef] [PubMed]
3. Eleftheriou, C.; Cesca, F.; Maragliano, L.; Benfenati, F.; Maya-Vetencourt, J.F. Optogenetic modulation of intracellular signalling and transcription: Focus on neuronal plasticity. *J. Exp. Neurosci.* **2017**, *11*. [CrossRef] [PubMed]
4. Deisseroth, K.; Feng, G.; Majewska, A.K.; Miesenböck, G.; Ting, A.; Schnitzer, M.J. Next-generation optical technologies for illuminating genetically targeted brain circuits. *J. Neurosci.* **2006**, *26*, 10380–10386. [CrossRef]
5. Boyden, E.S.; Zhang, F.; Bamberg, E.; Nagel, G.; Deisseroth, K. Millisecond-timescale, genetically targeted optical control of neural activity. *Nat. Neurosci.* **2005**, *8*, 1263–1268. [CrossRef]
6. Nagel, G.; Ollig, D.; Fuhrmann, M.; Kateriya, S.; Musti, A.M.; Bamberg, E.; Hegemann, P. Channelrhodopsin-1: A light-gated proton channel in green algae. *Science* **2002**, *296*, 2395–2398. [CrossRef]
7. Deisseroth, K.; Hegemann, P. The form and function of channelrhodopsin. *Science* **2017**, *357*, eaan5544. [CrossRef]
8. Deisseroth, K. Optogenetics: 10 years of microbial opsins in neuroscience. *Nat. Neurosci.* **2015**, *18*, 1213–1225. [CrossRef]
9. Berlin, S.; Isacoff, E.Y. Synapses in the spotlight with synthetic optogenetics. *EMBO Rep.* **2017**, *18*, 677–692. [CrossRef]
10. Berlin, S.; Isacoff, E.Y. *Optical Control of Glutamate Receptors of the NMDA-Kind in Mammalian Neurons, with the Use of Photoswitchable Ligands*; Springer Science+Business Media LLC: Berlin, Germany, 2017; Volume 130.
11. Ziegler, T.; Möglich, A. Photoreceptor engineering. *Front. Mol. Biosci.* **2015**, *2*. [CrossRef]
12. Paoletti, P.; Ellis-Davies, G.C.R.; Mourot, A. Optical control of neuronal ion channels and receptors. *Nat. Rev. Neurosci.* **2019**, *20*, 514–532. [CrossRef]
13. Folgering, J.H.A.; Kuiper, J.M.; de Vries, A.H.; Engberts, J.B.F.N.; Poolman, B. Lipid-mediated light activation of a mechanosensitive channel of large conductance. *Langmuir* **2004**, *20*, 6985–6987. [CrossRef] [PubMed]
14. Guerrero, L.; Smart, O.S.; Woolley, G.A.; Allemann, R.K. Photocontrol of DNA binding specificity of a miniature engrailed homeodomain. *J. Am. Chem. Soc.* **2005**, *127*, 15624–15629. [CrossRef] [PubMed]
15. Frank, J.A.; Franquelim, H.G.; Schwille, P.; Trauner, D. Optical control of lipid rafts with photoswitchable ceramides. *J. Am. Chem. Soc.* **2016**, *138*, 12981–12986. [CrossRef] [PubMed]
16. Habermacher, C.; Martz, A.; Calimet, N.; Lemoine, D.; Peverini, L.; Specht, A.; Cecchini, M.; Grutter, T. Photo-switchable tweezers illuminate pore-opening motions of an ATP-gated P2X ion channel. *eLife* **2016**, *5*. [CrossRef] [PubMed]
17. Kienzler, M.A.; Isacoff, E.Y. Precise modulation of neuronal activity with synthetic photoswitchable ligands. *Curr. Opin. Neurobiol.* **2017**, *45*, 202–209. [CrossRef]
18. Kramer, R.H.; Mourot, A.; Adesnik, H. Optogenetic pharmacology for control of native neuronal signaling proteins. *Nat. Neurosci.* **2013**, *16*, 816–823. [CrossRef]
19. Kaufman, H.; Vratsanos, S.M.; Erlanger, B.F. Photoregulation of an enzymic process by means of a light-sensitive ligand. *Science* **1968**, *162*, 1487–1489. [CrossRef]
20. Szymański, W.; Beierle, J.M.; Kistemaker, H.A.V.; Velema, W.A.; Feringa, B.L. Reversible photocontrol of biological systems by the incorporation of molecular photoswitches. *Chem. Rev.* **2013**, *113*, 6114–6178. [CrossRef]
21. Beharry, A.A.; Woolley, G.A. Azobenzene photoswitches for biomolecules. *Chem. Soc. Rev.* **2011**, *40*, 4422. [CrossRef]

22. Zhu, M.; Zhou, H. Azobenzene-based small molecular photoswitches for protein modulation. *Org. Biomol. Chem.* **2018**, *16*, 8434–8445. [CrossRef] [PubMed]
23. Hüll, K.; Morstein, J.; Trauner, D. In vivo photopharmacology. *Chem. Rev.* **2018**, *118*, 10710–10747. [CrossRef] [PubMed]
24. Fehrentz, T.; Schönberger, M.; Trauner, D. Optochemical Genetics. *Angew. Chem. Int. Ed.* **2011**, *50*, 12156–12182. [CrossRef] [PubMed]
25. Reiner, A.; Isacoff, E.Y. Photoswitching of cell surface receptors using tethered ligands. In *Photoswitching Protein*; Cambridge, S., Ed.; Springer: New York, NY, USA, 2014; Volume 1148, pp. 45–68. [CrossRef]
26. Berlin, S.; Szobota, S.; Reiner, A.; Carroll, E.C.; Kienzler, M.A.; Guyon, A.; Xiao, T.; Trauner, D.; Isacoff, E.Y. A family of photoswitchable NMDA receptors. *elife* **2016**, *5*. [CrossRef] [PubMed]
27. Broichhagen, J.; Damijonaitis, A.; Levitz, J.; Sokol, K.R.; Leippe, P.; Konrad, D.; Isacoff, E.Y.; Trauner, D. Orthogonal optical control of a G protein-coupled receptor with a SNAP-Tethered photochromic ligand. *ACS Cent. Sci.* **2015**, *1*, 383–393. [CrossRef] [PubMed]
28. Izquierdo-Serra, M.; Bautista-Barrufet, A.; Trapero, A.; Garrido-Charles, A.; Díaz-Tahoces, A.; Camarero, N.; Pittolo, S.; Valbuena, S.; Pérez-Jiménez, A.; Gay, M.; et al. Optical control of endogenous receptors and cellular excitability using targeted covalent photoswitches. *Nat. Commun.* **2016**, *7*. [CrossRef]
29. Donthamsetti, P.C.; Broichhagen, J.; Vyklicky, V.; Stanley, C.; Fu, Z.; Visel, M.; Levitz, J.L.; Javitch, J.A.; Trauner, D.; Isacoff, E.Y. Genetically Targeted Optical Control of an Endogenous G Protein-Coupled Receptor. *J. Am. Chem. Soc.* **2019**, *141*, 11522–11530. [CrossRef]
30. Banghart, M.; Borges, K.; Isacoff, E.; Trauner, D.; Kramer, R.H. Light-activated ion channels for remote control of neuronal firing. *Nat. Neurosci.* **2004**, *7*, 1381–1386. [CrossRef]
31. Fodje, M.N.; Al-Karadaghi, S. Occurrence, conformational features and amino acid propensities for the pi-helix. *Protein Eng.* **2002**, *15*, 353–358. [CrossRef]
32. Lester, H.A.; Krouse, M.E.; Nass, M.M.; Wassermann, N.H.; Erlanger, B.F. Light-activated drug confirms a mechanism of ion channel blockade. *Nature* **1979**, *280*, 509–510. [CrossRef]
33. McKenzie, C.K.; Sanchez-Romero, I.; Janovjak, H. Flipping the photoswitch: Ion channels under light control. In *Novel Chemical Tools to Study Ion Channel Biology*; Ahern, C., Pless, S., Eds.; Springer: New York, NY, USA, 2015; Volume 869, pp. 101–117. [CrossRef]
34. Kamei, T.; Akiyama, H.; Morii, H.; Tamaoki, N.; Uyeda, T.Q.P. Visible-Light photocontrol of (*E*)/(*Z*) isomerization of the 4-(Dimethylamino)azobenzene pseudo-nucleotide unit incorporated into an oligonucleotide and DNA hybridization in aqueous media. *Nucleosides Nucleotides Nucleic Acids* **2009**, *28*, 12–28. [CrossRef] [PubMed]
35. Ramakrishnan, P.; Maclean, M.; MacGregor, S.J.; Anderson, J.G.; Grant, M.H. Cytotoxic responses to 405nm light exposure in mammalian and bacterial cells: Involvement of reactive oxygen species. *Toxicol. In Vitro* **2016**, *33*, 54–62. [CrossRef] [PubMed]
36. Trigo, F.F.; Corrie, J.E.T.; Ogden, D. Laser photolysis of caged compounds at 405 nm: Photochemical advantages, localisation, phototoxicity and methods for calibration. *J. Neurosci. Methods* **2009**, *180*, 9–21. [CrossRef] [PubMed]
37. Zhang, H.; Salo, D.; Kim, D.M.; Komarov, S.; Tai, Y.-C.; Berezin, M.Y. Penetration depth of photons in biological tissues from hyperspectral imaging in shortwave infrared in transmission and reflection geometries. *J. Biomed. Opt.* **2016**, *21*. [CrossRef] [PubMed]
38. Weissleder, R. A clearer vision for in vivo imaging. *Nat. Biotechnol.* **2001**, *19*, 316–317. [CrossRef]
39. Wang, Z.; Hu, M.; Ai, X.; Zhang, Z.; Xing, B. Near-infrared manipulation of membrane ion channels via upconversion optogenetics. *Adv. Biosyst.* **2019**, *3*, 1800233. [CrossRef]
40. Ellis-Davies, G.C.R. Two-photon microscopy for chemical neuroscience. *ACS Chem. Neurosci.* **2011**, *2*, 185–197. [CrossRef]
41. Carroll, E.C.; Berlin, S.; Levitz, J.; Kienzler, M.A.; Yuan, Z.; Madsen, D.; Larsen, D.S.; Isacoff, E.Y. Two-photon brightness of azobenzene photoswitches designed for glutamate receptor optogenetics. *Proc. Natl. Acad. Sci. USA* **2015**, *112*, E776–E785. [CrossRef]
42. Hartley, G.S. The Cis-form of Azobenzene. *Nature* **1937**, *140*, 281. [CrossRef]
43. Shao, J.; Lei, Y.; Wen, Z.; Dou, Y.; Wang, Z. Nonadiabatic simulation study of photoisomerization of azobenzene: Detailed mechanism and load-resisting capacity. *J. Chem. Phys.* **2008**, *129*, 164111. [CrossRef]

44. Bandara, H.M.D.; Burdette, S.C. Photoisomerization in different classes of azobenzene. *Chem. Soc. Rev.* **2012**, *41*, 1809–1825. [CrossRef] [PubMed]
45. Hamm, P.; Ohline, S.M.; Zinth, W. Vibrational cooling after ultrafast photoisomerization of azobenzene measured by femtosecond infrared spectroscopy. *J. Chem. Phys.* **1997**, *106*, 519–529. [CrossRef]
46. Lednev, I.K.; Ye, T.Q.; Matousek, P.; Towrie, M.; Foggi, P.; Neuwahl, F.V.; Umapathy, S.; Hester, R.E.; Moore, J.N. Femtosecond time-resolved UV-visible absorption spectroscopy of trans-azobenzene: Dependence on excitation wavelength. *Chem. Phys. Lett.* **1998**, *290*, 68–74. [CrossRef]
47. Drobizhev, M.; Makarov, N.S.; Tillo, S.E.; Hughes, T.E.; Rebane, A. Two-photon absorption properties of fluorescent proteins. *Nat. Methods* **2011**, *8*, 393–399. [CrossRef]
48. Stoltzfus, C.R.; Barnett, L.M.; Drobizhev, M.; Wicks, G.; Mikhaylov, A.; Hughes, T.E.; Rebane, A. Two-photon directed evolution of green fluorescent proteins. *Sci. Rep.* **2015**, *5*, 1–9. [CrossRef]
49. Quick, M.; Dobryakov, A.L.; Gerecke, M.; Richter, C.; Berndt, F.; Ioffe, I.N.; Granovsky, A.A.; Mahrwald, R.; Ernsting, N.P.; Kovalenko, S.A. Photoisomerization dynamics and pathways of *trans*- and *cis*-azobenzene in solution from broadband femtosecond spectroscopies and calculations. *J. Phys. Chem. B* **2014**, *118*, 8756–8771. [CrossRef]
50. Moreno, J.; Gerecke, M.; Dobryakov, A.L.; Ioffe, I.N.; Granovsky, A.A.; Bleger, D.; Hecht, S.; Kovalenko, S.A. Two-photon-induced versus one-photon-induced isomerization dynamics of a bistable azobenzene derivative in solution. *J. Phys. Chem. B* **2015**, *119*, 12281–12288. [CrossRef]
51. Denk, W.; Strickler, J.; Webb, W. Two-photon laser scanning fluorescence microscopy. *Science* **1990**, *248*, 73–76. [CrossRef]
52. Benninger, R.K.P.; Piston, D.W. Two-photon excitation microscopy for the study of living cells and tissues. *Curr. Protoc. Cell. Biol.* **2013**, *49*, 4–11. [CrossRef]
53. Lee, G.H.; Moon, H.; Kim, H.; Lee, G.H.; Kwon, W.; Yoo, S.; Myung, D.; Yun, S.H.; Bao, Z.; Hahn, S.K. Multifunctional materials for implantable and wearable photonic healthcare devices. *Nat. Rev. Mater.* **2020**, 1–7. [CrossRef]
54. Rost, B.R.; Schneider-Warme, F.; Schmitz, D.; Hegemann, P. Optogenetic tools for subcellular applications in neuroscience. *Neuron* **2017**, *96*, 572–603. [CrossRef] [PubMed]
55. Oron, D.; Papagiakoumou, E.; Anselmi, F.; Emiliani, V. Chapter 7-Two-Photon Optogenetics. *Prog. Brain Res.* **2012**, *196*, 119–143. [CrossRef]
56. Chaigneau, E.; Ronzitti, E.; Gajowa, M.A.; Soler-Llavina, G.J.; Tanese, D.; Brureau, A.Y.; Papagiakoumou, E.; Zeng, H.; Emiliani, V. Two-photon holographic stimulation of ReaChR. *Front. Cell. Neurosci.* **2016**, *10*. [CrossRef] [PubMed]
57. Yang, W.; Carrillo-Reid, L.; Bando, Y.; Peterka, D.S.; Yuste, R. Simultaneous two-photon imaging and two-photon optogenetics of cortical circuits in three dimensions. *eLife* **2018**, *7*, e32671. [CrossRef] [PubMed]
58. Yizhar, O.; Fenno, L.E.; Prigge, M.; Schneider, F.; Davidson, T.J.; O'shea, D.J.; Sohal, V.S.; Goshen, I.; Finkelstein, J.; Paz, J.T.; et al. Neocortical excitation/inhibition balance in information processing and social dysfunction. *Nature* **2011**, *477*, 171–178. [CrossRef] [PubMed]
59. Ronzitti, E.; Ventalon, C.; Canepari, M.; Forget, B.C.; Papagiakoumou, E.; Emiliani, V. Recent advances in patterned photostimulation for optogenetics. *J. Opt.* **2017**, *19*, 113001. [CrossRef]
60. Forli, A.; Vecchia, D.; Binini, N.; Succol, F.; Bovetti, S.; Moretti, C.; Nespoli, F.; Mahn, M.; Baker, C.A.; Bolton, M.M.; et al. Two-photon bidirectional control and imaging of neuronal excitability with high spatial resolution in vivo. *Cell Rep.* **2018**, *22*, 3087–3098. [CrossRef] [PubMed]
61. Mohanty, S.K.; Reinscheid, R.K.; Liu, X.; Okamura, N.; Krasieva, T.B.; Berns, M.W. In-depth activation of channelrhodopsin 2-sensitized excitable cells with high spatial resolution using two-photon excitation with a near-infrared laser microbeam. *Biophys. J.* **2008**, *95*, 3916–3926. [CrossRef]
62. Ellis-Davies, G.C.R. Two-photon uncaging of glutamate. *Front. Synaptic Neurosci.* **2019**, *10*. [CrossRef]
63. Volgraf, M.; Gorostiza, P.; Numano, R.; Kramer, R.H.; Isacoff, E.Y.; Trauner, D. Allosteric control of an ionotropic glutamate receptor with an optical switch. *Nat. Chem. Biol.* **2006**, *2*, 47–52. [CrossRef]
64. Levitz, J.; Pantoja, C.; Gaub, B.; Janovjak, H.; Reiner, A.; Hoagland, A.; Schoppik, D.; Kane, B.; Stawski, P.; Schier, A.F.; et al. Optical control of metabotropic glutamate receptors. *Nat. Neurosci.* **2013**, *16*, 507–516. [CrossRef]

65. Izquierdo-Serra, M.; Gascón-Moya, M.; Hirtz, J.J.; Pittolo, S.; Poskanzer, K.E.; Ferrer, E.; Alibés, R.; Busqué, F.; Yuste, R.; Hernando, J.; et al. Two-photon neuronal and astrocytic stimulation with azobenzene-based photoswitches. *J. Am. Chem. Soc.* **2014**, *136*, 8693–8701. [CrossRef] [PubMed]
66. Gascón-Moya, M.; Pejoan, A.; Izquierdo-Serra, M.; Pittolo, S.; Cabré, G.; Hernando, J.; Alibés, R.; Gorostiza, P.; Busqué, F. An optimized glutamate receptor photoswitch with sensitized azobenzene isomerization. *J. Org. Chem.* **2015**, *80*, 9915–9925. [CrossRef] [PubMed]
67. Cabré, G.; Garrido-Charles, A.; Moreno, M.; Bosch, M.; Porta-de-la-Riva, M.; Krieg, M.; Gascón-Moya, M.; Camarero, N.; Gelabert, R.; Lluch, J.M.; et al. Rationally designed azobenzene photoswitches for efficient two-photon neuronal excitation. *Nat. Commun.* **2019**, *10*, 1–12. [CrossRef] [PubMed]
68. Rullo, A.; Reiner, A.; Reiter, A.; Trauner, D.; Isacoff, E.Y.; Woolley, G.A. Long wavelength optical control of glutamate receptor ion channels using a tetra-ortho-substituted azobenzene derivative. *Chem. Commun.* **2014**, *50*, 14613–14615. [CrossRef] [PubMed]
69. Laprell, L.; Repak, E.; Franckevicius, V.; Hartrampf, F.; Terhag, J.; Hollmann, M.; Sumser, M.; Rebola, N.; DiGregorio, D.A.; Trauner, D. Optical control of NMDA receptors with a diffusible photoswitch. *Nat. Commun.* **2015**, *6*. [CrossRef]
70. Riefolo, F.; Matera, C.; Garrido-Charles, A.; MJGomila, A.; Sortino, R.; Agnetta, L.; Claro, E.; Masgrau, R.; Holzgrabe, U.; Batlle, M.; et al. Optical control of cardiac function with a photoswitchable muscarinic agonist. *J. Am. Chem. Soc.* **2019**, *141*, 7628–7636. [CrossRef]
71. Pittolo, S.; Lee, H.; Lladó, A.; Tosi, S.; Bosch, M.; Bardia, L.; Gómez-Santacana, X.; Llebaria, A.; Soriano, E.; Colombelli, J.; et al. Reversible silencing of endogenous receptors in intact brain tissue using 2-photon pharmacology. *Proc. Natl. Acad. Sci. USA* **2019**, *116*, 13680–13689. [CrossRef]
72. Cabré, G.; Garrido-Charles, A.; González-Lafont, A.; Moormann, W.; Langbehn, D.; Egea, D.; Lluch, J.M.; Herges, R.; Alibés, R.; Busqué, F.; et al. Synthetic photoswitchable neurotransmitters based on bridged azobenzenes. *Org. Lett.* **2019**, *21*, 3780–3784. [CrossRef]
73. Passlick, S.; Richers, M.T.; Ellis-Davies, G.C.R. Thermodynamically stable, photoreversible pharmacology in neurons with one- and two-photon excitation. *Angew. Chem. Int. Ed.* **2018**, *57*, 12554–12557. [CrossRef]
74. Dong, M.; Babalhavaeji, A.; Samanta, S.; Beharry, A.A.; Woolley, G.A. Red-shifting azobenzene photoswitches for in vivo use. *Acc. Chem. Res.* **2015**, *48*, 2662–2670. [CrossRef]
75. Kienzler, M.A.; Reiner, A.; Trautman, E.; Yoo, S.; Trauner, D.; Isacoff, E.Y. A red-shifted, fast-relaxing azobenzene photoswitch for visible light control of an ionotropic glutamate receptor. *J. Am. Chem. Soc.* **2013**, *135*, 17683–17686. [CrossRef]
76. Hansen, M.J.; Lerch, M.M.; Szymanski, W.; Feringa, B.L. Direct and versatile synthesis of red-shifted azobenzenes. *Angew. Chem. Int. Ed.* **2016**, *55*, 13514–13518. [CrossRef]
77. Mourot, A.; Kienzler, M.A.; Banghart, M.R.; Fehrentz, T.; Huber, F.M.; Stein, M.; Kramer, R.H.; Trauner, D. Tuning photochromic ion channel blockers. *ACS Chem. Neurosci.* **2011**, *2*, 536–543. [CrossRef]
78. Carmi, I.; De Battista, M.; Maddalena, L.; Carroll, E.C.; Kienzler, M.A.; Berlin, S. Holographic two-photon activation for synthetic optogenetics. *Nat. Protoc.* **2018**, *14*, 864, under final review. [CrossRef]
79. Kim, H.M.; Cho, B.R. Two-photon probes for intracellular free metal ions, acidic vesicles, and lipid rafts in live tissues. *Acc. Chem. Res.* **2009**, *42*, 863–872. [CrossRef]
80. Dong, M.; Babalhavaeji, A.; Hansen, M.J.; Kálmán, L.; Woolley, G.A. Red, far-red, and near infrared photoswitches based on azonium ions. *Chem. Commun.* **2015**, *51*, 12981–12984. [CrossRef]
81. Sell, H.; Näther, C.; Herges, R. Amino-substituted diazocines as pincer-type photochromic switches. *Beilstein J. Org. Chem.* **2013**, *9*, 1–7. [CrossRef]
82. Thapaliya, E.R.; Zhao, J.; Ellis-Davies, G.C.R. Locked-azobenzene: testing the scope of a unique photoswitchable scaffold for cell physiology. *ACS Chem. Neurosci.* **2019**, *10*, 2481–2488. [CrossRef]
83. Hartrampf, F.W.W.; Barber, D.M.; Gottschling, K.; Leippe, P.; Hollmann, M.; Trauner, D. Development of a photoswitchable antagonist of NMDA receptors. *Tetrahedron* **2017**, *73*, 4905–4912. [CrossRef]
84. GÓmez-Santacana, X.; Pittolo, S.; Rovira, X.; Lopez, M.; Zussy, C.; Dalton, J.A.; Faucherre, A.; Jopling, C.; Pin, J.P.; Ciruela, F.; et al. Illuminating phenylazopyridines to photoswitch metabotropic glutamate receptors: from the flask to the animals. *ACS Cent. Sci.* **2017**, *3*, 81–91. [CrossRef]
85. Volgraf, M.; Gorostiza, P.; Szobota, S.; Helix, M.R.; Isacoff, E.Y.; Trauner, D. Reversibly caged glutamate: a photochromic agonist of ionotropic glutamate receptors. *J. Am. Chem. Soc.* **2007**, *129*, 260–261. [CrossRef]

86. Jiang, C.; Li, H.T.; Zhou, Y.M.; Wang, X.; Wang, L.; Liu, Z.Q. Cardiac optogenetics: A novel approach to cardiovascular disease therapy. *EP Eur.* **2017**, *20*, 1741–1749. [CrossRef]
87. Gepstein, L.; Gruber, A. Optogenetic neuromodulation of the heart. *J. Am. Coll. Cardiol.* **2017**, *70*, 2791–2794. [CrossRef]
88. Boyle, P.M.; Karathanos, T.V.; Trayanova, N.A. Cardiac optogenetics: 2018. *JACC Clin. Electrophys.* **2018**, *4*, 155–167. [CrossRef]
89. Tochitsky, I.; Kienzler, M.A.; Isacoff, E.; Kramer, R.H. Restoring vision to the blind with chemical photoswitches. *Chem. Rev.* **2018**, *118*, 10748–10773. [CrossRef]
90. Gaub, B.M.; Berry, M.H.; Visel, M.; Holt, A.; Isacoff, E.Y.; Flannery, J.G. Optogenetic retinal gene therapy with the light gated GPCR vertebrate rhodopsin. In *Retinal Gene Therapy*; Boon, C.J.F., Wijnholds, J., Eds.; Springer: New York, NY, USA, 2018; Volume 1715, pp. 177–189. [CrossRef]
91. Berry, M.H.; Holt, A.; Levitz, J.; Broichhagen, J.; Gaub, B.M.; Visel, M.; Stanley, C.; Aghi, K.; Kim, Y.J.; Cao, K.; et al. Restoration of patterned vision with an engineered photoactivatable G protein-coupled receptor. *Nat. Commun.* **2017**, *8*. [CrossRef]
92. Hüll, K.; Benster, T.; Manookin, M.B.; Trauner, D.; Van Gelder, R.N.; Laprell, L. Photopharmacologic vision restoration reduces pathological rhythmic field potentials in blind mouse retina. *Sci. Rep.* **2019**, *9*, 1–12. [CrossRef]
93. Samia, S.; Berlin, S. Bringing synthetic optogenetics to the clinic. *J. Cell Signal.* **2017**, *2*. [CrossRef]
94. Donthamsetti, P.C.; Winter, N.; Schönberger, M.; Levitz, J.; Stanley, C.; Javitch, J.A.; Isacoff, E.Y.; Trauner, D. Optical control of dopamine receptors using a photoswitchable tethered inverse agonist. *J. Am. Chem. Soc.* **2017**, *139*, 18522–18535. [CrossRef]
95. Borowiak, M.; Nahaboo, W.; Reynders, M.; Nekolla, K.; Jalinot, P.; Hasserodt, J.; Rehberg, M.; Delattre, M.; Zahler, S.; Vollmar, A.; et al. Photoswitchable inhibitors of microtubule dynamics optically control mitosis and cell death. *Cell* **2015**, *162*, 403–411. [CrossRef]
96. Cela, E.; Sjöström, P.J. Novel optogenetic approaches in epilepsy research. *Front. Neurosci.* **2019**, *13*. [CrossRef] [PubMed]
97. Gradinaru, V.; Zhang, F.; Ramakrishnan, C.; Mattis, J.; Prakash, R.; Diester, I.; Goshen, I.; Thompson, K.R.; Deisseroth, K. Molecular and cellular approaches for diversifying and extending optogenetics. *Cell* **2010**, *141*, 154–165. [CrossRef] [PubMed]
98. Fenno, L.; Yizhar, O.; Deisseroth, K. The development and application of optogenetics. *Annu. Rev. Neurosci.* **2011**, *34*, 389–412. [CrossRef] [PubMed]
99. Chen, I.W.; Ronzitti, E.; Lee, B.R.; Daigle, T.L.; Dalkara, D.; Zeng, H.; Emiliani, V.; Papagiakoumou, E. In vivo submillisecond two-photon optogenetics with temporally focused patterned light. *J. Neurosci.* **2019**, *39*, 3484–3497. [CrossRef]

MDPI
St. Alban-Anlage 66
4052 Basel
Switzerland
Tel. +41 61 683 77 34
Fax +41 61 302 89 18
www.mdpi.com

Applied Sciences Editorial Office
E-mail: applsci@mdpi.com
www.mdpi.com/journal/applsci

www.ingramcontent.com/pod-product-compliance
Lightning Source LLC
LaVergne TN
LVHW070843170726
843515LV00005B/564